INNOVATION: ADAPTATION AND GROWTH

An International Perspective

INNOVATION: ADAPTATION AND GROWTH

An International Perspective

Proceedings of the Fifth International Conference on Product Innovation Management, Brighton, United Kingdom, August 24–26, 1987

(Run in conjunction with the International Society for Product Innovation Management, Brighton Business School, Brighton Polytechnic and the Science Policy Research Unit, University of Sussex)

edited by

Roy Rothwell

Professor, Science Policy Research Unit, University of Sussex, Brighton, United Kingdom

and

John Bessant

Professor, Department of Research, Brighton Business School, Brighton Polytechnic, Brighton, United Kingdom

ELSEVIER

Amsterdam — Oxford — New York — Tokyo 1987

ELSEVIER SCIENCE PUBLISHERS B.V.
Sara Burgerhartstraat 25
P.O. Box 211, 1000 AE Amsterdam, The Netherlands

Distributors for the United States and Canada:

ELSEVIER SCIENCE PUBLISHING COMPANY INC.
52, Vanderbilt Avenue
New York, NY 10017, U.S.A.

ISBN 0-444-42861-5

Printed in The Netherlands

FOREWORD

Public policies in the advanced market economies and, increasingly, in the newly industrializing nations, are placing more and more emphasis on the generation and diffusion of technological innovations. Technology is recognized as a crucial factor determining manufacturing productivity increase and non-price competitiveness in manufactures. At the same time the rapidly expanding service sector is increasingly utilizing technology-based products flowing from developments in information technology: this is well exemplified in the recent so called "big bang" - the computerization of financial services in the City of London.

The creation and adoption of technological innovations require companies to formulate and implement appropriate strategies. Simple portfolio techniques are no longer sufficient and current high rates of technological change mandate that "technology strategies" become a core feature of overall corporate strategies.

The Conference addresses four main issues. Firstly, it is devoted to strategic issues, focusing on strategies that relate specifically to firms' innovatory activities. Of particular interest here are strategies that involve technology users as a key element in producers' innovation processes. At the same time, a clear distinction is made between strategies for the creation of innovations and those directed towards innovation adoption and adaptation. In general terms, the former are product-market strategies while the latter are largely manufacturing strategies although the two are not, of course, unrelated.

Secondly, the Conference focuses on public innovation policies. While it is generally acknowledged that innovatory success or failure depends mainly on the abilities and propensities of firms' managements, it is nevertheless also recognized that public policies can enhance firms' innovatory performance. Public policy initiatives can remove barriers to innovation, provide crucial inputs (technology, finance, skills) and create an overall climate which is conducive to risky innovatory activities at the level of the individual enterprise. Such policies are particularly effective when they are designed to complement firms' own innovation strategies. A variety of public policy issues are tackled, ranging from cooperative research to technology transfer and manpower planning. The creation of science parks (a long-established phenomenon in the United States) is now the fastest growing technology-stimulating infrastructural mechanism being employed throughout Europe.

Thirdly, the Conference focuses entirely on issues relating to the establishment and management of science parks.

Figure 1 shows the pattern of evolution of public policies for the stimulation of scientific and technological advance and industrial innovation, from 1950 to date. Perhaps the most marked trend has been the rapidly growing emphasis on assisting innovatory developments in small firms in general since the mid-1970s and the focus on new technology-based firms (NTBFs) in particular during the 1980s. Small firms, it is felt, can play an especially important role in technological innovation and in employment creation; NTBFs can have an important catalytic effect in stimulating industrial structural shifts towards new technology-based, more knowledge-intensive sectors and product groups.

Finally, the Conference reflects this growing policy pre-occupation with the small firm sector and is devoted entirely to innovation and management practices in small firms. It covers such issues as innovation strategies, inter-company linkages and the characteristics of high-growth technology-based small firms in Europe.

An indispensible task for public and private policy makers alike is the proper management of the processes of innovation generation and diffusion. As the pace of technological change accelerates and product cycles shorten in key sectors, the ability to formulate and implement imaginative and dynamic technology

Figure 1

EVOLUTION OF PUBLIC RESEARCH AND TECHNOLOGY DEVELOPMENT POLICIES

1950's and 1960's	SCIENCE POLICY - SCIENTIFIC EDUCATION - UNIVERSITY RESEARCH - BASIC RESEARCH IN GOVERNMENT LABS.	INDUSTRIAL POLICY - GRANTS FOR R&D - EQUIPMENT GRANTS - INDUSTRIAL RESTRUCTURING - SUPPORT FOR COLLECTIVE INDUSTRIAL RESEARCH - TECHNICAL EDUCATION AND TRAINING ETC.	FIRM SIZE EMPHASIS - EMPHASIS ON LARGE FIRMS AND INDUSTRIAL AGGLOMERATION - CREATING NATIONAL FLAGSHIP COMPANIES, E.G. IN COMPUTERS	LITTLE VENTURE CAPITAL AVAILABLE
	LITTLE COORDINATION OR ACTIVE COLLABORATION			
MID-1970's	INNOVATION POLICY - GRANTS FOR 'INNOVATION' - INVOLVING COLLECTIVE RESEARCH INSTITUTES IN PRODUCT DEVELOPMENT - INNOVATION-STIMULATING PUBLIC PROCUREMENT ETC. (INCREASED INTER-DEPARTMENTAL COORDINATION)		- INCREASING EMPHASIS ON SMALL FIRMS	
1980's INCREASED EMPHASIS ON SCIENCE POLICY	STRATEGIC INNOVATION POLICY (TECHNOLOGY POLICY) - SELECTION AND SUPPORT OF GENERIC TECHNOLOGIES AND HIGH-TECHNOLOGY PRODUCT GROUPS - INTERNATIONAL (EC) COLLABORATION (INTER-DEPARTMENTAL INITIATIVES)		- EMPHASIS ON NTBFs AND INTER-FIRM COLLABORATION	INCREASE IN THE AVAILABILITY OF VENTURE CAPITAL

SOURCE: R. Rothwell and W. Zegveld, Reindustrialization and Technology, Longman, Harlow, U.K., 1985,
and
R. Rothwell, "Preface" to Technology Policy, special issue of Technovation, Vol. 5, February 1987.

strategies at both the national and private enterprise levels has become a major factor determining economic growth and corporate survival and profitability. The papers presented to the Fifth International ISPIM Conference will be of considerable assistance to strategists at all levels who are concerned with technological change and its relationship to economic growth, productivity increase and enhanced international competitiveness.

Prof. Roy Rothwell
Prof. John Bessant

CONTENTS

Innovation: Adaptation and Growth, edited by R. Rothwell and J. Bessant
Elsevier Science Publishers B.V., Amsterdam, 1987 — Printed in The Netherlands

THE ROLE OF THE USER IN PRODUCT INNOVATION

Knut Holt

Division of Organization and Work Science, University of Trondheim, Norwegian Institute of Technology (Norway)

SUMMARY

The provision of information about user needs is often limited to the experts who participate in the product innovation process. However, usually they are not good user representatives. Therefore, increasingly the user is being involved. It may be through informal contacts with users, by users contacting the company, .and by company staff with user contacts. One may also generate user information by applying methods such as statistical sampling of users, observation of users, simulation of user situations, brainstorming/brainwriting with user participation, user delphi, user employment, negative brainstorming with user participation, questioning of users, user projects, and product development by users. The practical application of the need assessment methods depends on company factors such as corporate culture/management philosophy, attitude toward risk, technology strategy, size of the company, and resources; individual factors such as attitude/motivation, fantasy, knowledge/education, and experience; product factors such as importance of the product, degree of novelty, degree of customation, and place in the life cycle; project factors such as cost of the need assessment study, place of the study in the project cycle, and confidentialiy; external factors such as practice of other companies, particularly competitors, type of market, competence of users, access to users, distance to users, and user stakeholders. Some companies are using one method all the time, others are using several methods, and there are companies that have an eclectic approach, where the choice of methods depends on the situation.

INTRODUCTION

During the last 20 years the user has increasingly been brought into focus in connection with product innovation activities. A number of empirical studies (refs. 1-9) show that most successful new products are initiated by information about user needs, often referred to as "need pull". A study from the USA resulted in the best-seller "The Search of Excellence" (ref. 10), which brought the message out to the whole industrial world. One of the major conclusions is that in too many firms the users are considered a nuisance despite all lip service given to market-orientation. The excellent firms behave in a different way. They consider it to be of vital importance to keep close contact with the user in order to get proper information about his needs and preferences.

The situation in Europe appears to be the same as in the USA. Studies in German, Italian and Norwegian firms show that only few companies pay attention to user needs (ref. 11). Need information in most companies originates by chance, through informal contacts, or by gifted individuals who are sensitive to human needs.

The purpose of this paper is to contribute to more systematic need assesment (SNA) by discussing how the user can be involved in product innovation processes.

NEED ASSESSMENT METHODS

The link between the producer and the user can be arranged in many ways as indicated in Fig. 1.

At the one extreme we have the expert approach. The user is here represented by an engineer, a marketer or an industrial designer who puts himself in the place of the user and attempts to visualize his needs. In most cases this is not a good solution. The engineer is a bad represenative of the user, and the marketer is even worse. The industrial designer perhaps falls some place between.

At the other extreme we find the user who is developing the product himself. This does not occur only in the hobby field. Studies by von Hippel (ref. 12) give examples of products, mostly production equipment and instruments, that have been developed and even produced, by users.

Between the extremes we have several methods where the user is more or less involved. The majority of them are best suited at the production stage, i.e. for improvement of existing products. Although the distinction is not clear, one can divide the methods into two groups.

Utilization of existing information

In this group we find information that is located within the company, often in a rather unorganized form. We have here four methods, all providing information more or less through chance ocurrences. They represent a relatively cheap way of determining user needs, but mostly they lead only to minor improvements. The quality of the result can be improved by stimulating those involved to participate actively, and by developing procedures that channels the information to those who will benefit from it in their product innovation activities.

	METHODS \ STAGES	IDEA	CON-CEPT	PRO-TO-TYPE	TEST RUN	PRO-DUC-TION
EXISTING INF.	1. EXPERT REPRESENTATION	X	X	X	X	X
	2. INFORMAL USER CONTACTS					X
	3. INFORMATION FROM CUSTOMERS					X
	4. INFORMATION FROM STAFF WITH CUSTOMER CONTACTS					X
GENERATION OF NEW INFORMATION	5. STATISTICAL SAMPLING OF USERS					X
	6. OBSERVATION OF USERS					X
	7. SIMULATION OF USER SITUATIONS					X
	8. BRAINSTORMING/BRAINWRITING WITH USER PARTICIPATION					X
	9. USER DELPHI					X
	10. USER EMPLOYMENT		X	X	X	X
	11. NEGATIVE BRAINSTORMING WITH USER PARTICIPATION		X	X	X	X
	12. QUESTIONING OF USERS		X	X	X	X
	13. USER CLUBS	X	X	X	X	X
	14. USER PROJECTS	X	X	X	X	X
	15. USER DEVELOPMENT	X	X	X	X	X

Fig. 1. User involvement in the product innovation process

Generation of new information

This approach requires more effort. One has to plan and implement special activities, but the information is usually more complete and reliable.

Traditional market research based on statistical sampling may lead to minor improvements. However, it does not give that understanding of the user and his problems that is essential for perceiving needs which could lead to radically new products. Specific SNA methods are then required.

The development of entirely new products is most often based on visions. In such situations it is not possible to start the process by defining the problem in terms of the needs of those involved. On the contrary, the problem will only be presented in a general way, or it will have the character of a dream, a vision, of an intuitive sense of what must be done in order to reach a desired future state. A person with visions has the capability of anticipating future conditions and of perceiving opportunities and needs ahead of anybody else, and of communicating the vision in a simple, easily understood form to those concerned. Few people are able to do this, but the application of creative techniques with user participation can be of help. One example of a successful application is a lock system for hotel rooms where the traditional key is replaced by a punch-card in plastics. For each guest a new card is made with a different pattern. If the key is lost, a new and different combination can immediately be made. This concept was developed by means of brainstorming. In the session participated user stakeholders such as representatives for the guests, the lobby staff, the security people, the maintenance crew, the local management, and the corporate management. Based on experience from existing lock systems were indicated wishes and requirements to an ideal system. By grouping and giving priority to the various requirements one was able to specify the major functions of the new system. One could then proceed and generate ideas and select a feasible solution concept.

APPLICATION

A number of factors will influence the application of the methods for systematic need assessment. Several of them are shown in Fig. 2.

Our current knowledge is limited with regard to relevant factors, their relative importance, and to what degree they are interrelated. Many of them have not been defined operationally, and others have not been measured in a satisfactory way. Therefore, in order to comment upon them we have to rely on "educated guesses" founded on personal knowledge and experience.

- COMPANY FACTORS; CORPORATE CULTURE/MANAGEMENT PHILOSOPHY, ATTITUDE TOWARD RISK, TECHNOLOGY STRATEGY, SIZE, RESOURCES
- INDIVIDUAL FACTORS; ATTITUDES/MOTIVATION, FANTASY, KNOWLEDGE/EDUCATION, EXPERIENCE
- PRODUCT FACTORS; IMPORTANCE OF THE PRODUCT, DEGREE OF NOVELTY, DEGREE OF CUSTOMATION, PLACE IN THE LIFE CYCLE
- PROJECT FACTORS; COST OF THE SNA STUDY, PLACE OF THE STUDY IN THE PROJECT CYCLE, CONFIDENTIALITY
- EXTERNAL FACTORS; PRACTICE OF OTHER COMPANIES, PARTICULARLY COMPETITORS, TYPE OF MARKET, COMPETENCE OF USERS, ACCESS TO USERS, DISTANCE TO USERS, USER STAKEHOLDERS

Fig. 2. Factors influencing the application of SNA methods

Company factors

The corporate culture is one of the most important factors. In a company with a strong culture, rooted in a market-oriented management philosophy, one is concerned with providing people what they want, rather than to persuade them to buy what the engineers think they want or find it convenient to make. One has a good understanding of the market and pays attention to the problems and the needs of the user.

Quite different is the situation in the technology driven company that attempts to obtain technological leadership by developing new products with superior functional performance. The ideas tend to be of a visionary nature. As soon as a solution is found that works, one quickly brings it out to the market in order to reap the benefits by being first. However, often it does not succeed, first of all due to lack of market understanding. This is particularly true for many of the high-tech companies, that according to Drucker (ref. 13) are being mismanaged. The founders are still rather nineteenth-century inventors than twentieth-century innovators, believing in Benjamin Franklin's dictum: "If you invent a better mousetrap, the world will beat a path to your door". However, it does not occur to them to ask what makes a mousetrap better. They tend to be infatuated with their own technology, of believing that quality means what is technically sophisticated rather than what gives value to the user. This gives the company with a "follow-the-leader" strategy an opportunity. By looking at the product from the point of view of the user it can exploit the results of the pioneer, for example by adding new features or by providing different versions for different segments of the market.

An overview of the situation is given in Fig. 3. At the upper left corner is indicated a rather unusual case, a highly innovative company that at the same time has a good understanding of the market. The opposite corner is represented by the laggard who neglects both the technology and the market. This is no unusual situation. The adoption of a new technology, marketing concept, or management technique, is as, indicated in Fig. 4, a time-consuming process. Even in those cases where the benefits are obvious, there are many firms that neglect their market and operate with a technology far below state of the art.

MARKET UNDERSTANDING / TECHNOLOGY LEVEL	GOOD	AVERAGE	BAD
CUTTING EDGE (LEADERSHIP)	STAR	SUCCESS	DOUBT
STATE OF THE ART (FOLLOW THE LEADER - ME TOO)	SUCCESS	DOUBT	FAILURE
OBSOLATE	DOUBT	FAILURE	DISASTER

Fig. 3. The technology/market matrix

In between the extremes we find, as shown in Fig. 3, various combinations. For example, it is suggested that a successful product can be obtained by superior technology and average market understanding, or by state of the art technology and good market knowledge.

The studies referred to above (refs. 10, 11) indicate that most companies are not market driven. Some are technology driven, but most appear to be "by chance" driven, i.e. dependent on the initiative, background, and capability of individuals. A study by Juhler (ref. 14) points in the same direction. Most need information is provided by methods requiring as little resources as possible, e.g. informal contacts with dealers. If the information differs from the opinions of those involved in the product innovation process, it is often neglected.

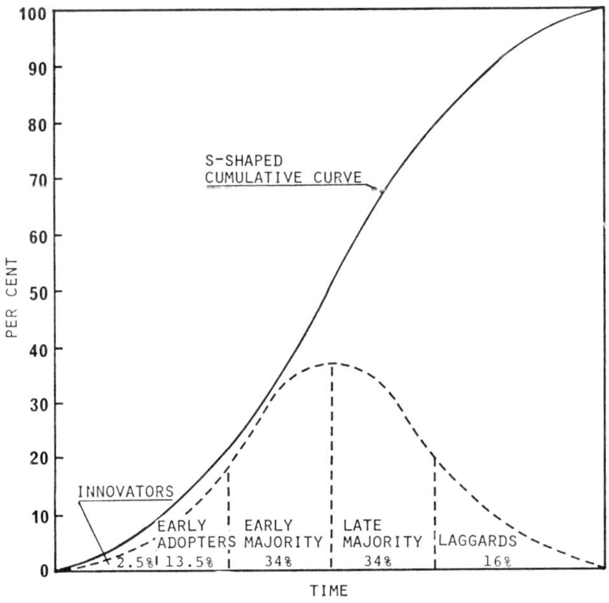

Fig. 4. Theoretical diffusion curve and adopted distribution

Individual factors

The attitude and motivation of the individual is an important factor. Even in a technology-oriented culture an engineer or marketer may successfully apply SNA methods. However, he must expect resistance and be prepared to demonstrate the value of the methods by his own behaviour.

Creative individuals may use their fantasy and develop original approaches for providing need related information. There are numerous ways of applying SNA methods. Take for example user questioning. One may use structured and unstructured interviews, in-depth interviews, group interviews, telephone surveys, mail surveys, user meetings, etc. One may also be more sophisticated. An example is a producer of cars who shows fullscale anonymous models, made in clay, wood or plastics, to a large number of users. Their reactions are listed by themselves on special forms, or by special interviewers who follow and record comments by means of a tape recorder. A future development may be to use video technology. Each participant sit at a screen and study the models. By means of precorded sequences comparisons are made with competing products. A further refinement may be done to expose the users to holographic images from interference patterns created by a split beam of laser light.

Product factors

The degree of novelty is an important product factor. Where the objective is to improve an existing product one may start with user information provided by means of informal user contact, customer information, and staff information. The material may be supplemented with in-depth studies by means of user questioning, user projects, and negative brainstorming. The latter is a variation of classical brainstorming. In a session, where users are participating, the energy is concentrated on finding faults, fault-possibilities, weaknesses, and lacking or unfulfilled functions in the product (negative brainstorming may also be used to study concepts and prototypes). The session may last up to two hours. It usually results in 30-50, or more, possible defects associated with the product itself, the environment where it is used, and the persons using it. After the session is completed follows evaluation of the proposals; it may take up to four hours. Negative brainstorming requires a critical attitude. As it is difficult for those who have participated in the project to change attitude, one may, in addition to user representatives, include a few people from the company staff who have not been involed. The method should be used with care; a too strong focus on the negative aspects may kill the idea.

When one is faced with an entirely new product of a visionary nature, the user is not able to recognize the need-fulfilling potential of the proposed technology which he does not, or only vaguely, understands. A well designed concept test may be useful, but often the first need assessment is undertaken at the prototype stage, or even later. The reason for this may be a wish to keep the new product secret as long as possible. One may also be so preoccupied with developing the technology and bringing the new product to the market, that there is no time for provision of market information.

Project factors

Among the project factors can be mentioned the place of the study in the project. It is rather common to involve users in the testing of prototypes. There are also firms that make a conscious effort of obtaining the reactions of the user before starting to develop the technology. One example is found at a producer of transportation equipment for assembly lines. First the concept was discussed with knowledgeable users in twelve companies. Then a written concept test was mailed to 1,000 firms. Finally, a physical model was presented and discussed with invited participants from eleven firms, all potential users with experience from assembly operations.

External factors

The type of market has considerble influence on the selection of SNA methods. In the industrial market direct cooperation with lead users, for example in terms of joint projects, can provide valuable information. This is not feasible in the consumer market; here one has to rely more on questioning approaches.

NEED ASSESSMENT PRACTICE

In order to provide reliable need information one should study the various methods and evaluate them by means of an analysis of the factors presented above. This will provide a basis for the selection and practical application of the SNA methods.

The mono-dimensional approach

Some companies select and use one method for all need assessment purposes. Companies that focus on improvement of existing products may for example use negative brainstorming in connection with concepts, prototypes, test runs, and regular products. In companies where the focus is on the development of new products of a pioneering nature, brainstorming or brainwriting may be useful. Joint projects with users may also give good results. One example is a producer of medical instruments. The products are developed in close cooperation with lead-users (ref. 15). In a joint project the current situation of the user is recorded and studied by means of system analysis including activity diagrams, work procedures, and information flow. Prominent doctors, often from several countries, participate during the whole project and assist in the market introduction of the product.

An important advantage with the mono-dimensional approach is that those involved become very familiar with the method. On the other hand, there is the danger of becoming so one-sided and technique-oriented that every problem is handled in the same way, whether justified or not.

The multi-dimensional approach

Most companies use several SNA methods. One example is a producer of fishing equipment. Before starting the development of a new product, current fishing methods are studied in close cooperation with fishermen. By questioning them, by observing them at work, and by studying their behaviour on shore and on board, one is able to collect valuable user information. In connection with improvement of existing products, information is provided by user observation, user questioning, analysis of complaints and service reports, employment of people with user experience, and analysis of information from the marketing organization.

The major advantage of using several need assessment methods is that one can select the approach that is best fitted for the problem at hand, provided that the choice is based on a thorough review of pros and cons of available alternatives. In many ways the situation is here like that of the experienced craftsman. He knows the tools of his trade, and how to apply them. Look at a carpenter; depending on the task, he selects the tools that are best fitted for the job. In a similar way those involved in the product innovation process may master several need assessment methods.

By using the "tool kit" approach one is likely to get better and more comprehensive information, and thereby a broader basis for formulation of need specification.

The eclectic approach

Here one does not have a fixed practice, but selects and applies what is considered the best elements of the various approaches. In some cases it may be one method, in other cases several. This gives a flexible approach determined by the type of problem, the resources available, and the preference of the person making the study.

Whatever approach is used, the study will end up with a long list with problems and needs, which is converted into a need specification. First, they must be clustered, i.e. arranged in groups related to performance, operation, maintenance, safety, status, appearence, etc. The various groups are then weighted, for example by questioning of users.

CONCLUSION

Marketing has a major role to play in product innovation. Modern marketing starts with the market, not with the product. The focus is on the needs of the user - "attempting to satisfy human needs without first assessing them accurately is like attempting to deliver a package without knowing the address". Yet, in many companies the user is neglected, and even forgotten. It does not have to be like this. An array of methods are available for the systematic assessment of user needs. In several of them the users are involved. However, the users may not always know what they want. It is for example difficult to recognize the benefits of break-through products that require radical changes in behaviour.

There are many ways of providing user information. Active need experience can be a useful approach. One then studies what is unsatisfactory in a user situation by actually working in a relevant environment for a certain period of time. Other sources are government information, competitor information, trade fairs, literature and experts. The real challenge to the participants in the product innovation process is to select an approach that will provide a good understanding of the problems and the needs of the user. In such considerations methods involving the user will have an important place.

REFERENCES

1 N.R. Baker, J. Siegman and A.H. Rubinstein, The Effects of Perceived Needs and Means of the Generation of Ideas for Industrial Research and Development Processes, IEEE Transactions on Engineering Management, December (1967) 156-163.

2 S. Myers and D. Marquis, Successful Industrial Innovations, Superintendant of Documents, Washington DC, 1969, 177 p.

3 A. Gerstenfeld, Effective Management of Research and Development, Addison-Wesley, Reading, 1970, 150 p.

4 J.M. Utterback, The Process of Innovation - a Study of the Origination and Development of Ideas for Scientific Instruments, IEEE Transactions on Engineering Management, November (1971) 124-131.

5 J. Langrish, M. Gibbson, W.G. Evans and F.R. Jevons, Wealth of Knowledge, Macmillan, Edinburgh, 1972, 477 p.

6 D. Robertson, The Marketing Factor in Successful Industrial Innovations, Industrial Marketing Management No. 4 (1973) 369-374.

7 R. Rothwell, C. Freeman, A. Horsley, V.T.P. Jervis, A.B. Robertson and J. Towsend, Sappho Updated - Project Sappho Phase II, Research Policy, No. 3 (1974) 258-291.

8 TNO, Innovatieprocessen in de Nederlandse industrie, TNO, Apeldoorn, 1974, 158 p.

9 H. Kulvik, Factors Underlying the Success or Failure of New Products, Laboratories of Industrial Economics and Industrial Psychology, Helsinki University of Technology, Otaniemi, 1977, 49 p.

10 T.J. Peters and R.H. Waterman, The Search of Excellence, Lessons from America's Best Run Companies, Harper & Row, New York, 1982, 360 p.

11 K. Holt, H. Geschka and G. Peterlongo, Need Assessment - a Key to User-Oriented Product Innovation, John Wiley, Chichester, 1984, 187 p.

12 E. von Hippel, Get New Products from Customers, Harvard Business Review, March-April (1982) 117-122.

13 P. Drucker, Innovation and Entrepreneurship, Harper & Row, New York, 1986, 277 p.

14 H. Juhler, Markedsorientert produktutvikling, Master thesis, University of Trondheim, Norwegian Institute of Technology, Trondheim, 1986, 150 p.

15 J.R. Olsson, The Market-Leader Method in User Oriented Product Development, Unpublished Manuscript, 1986, 10 p.

Innovation: Adaptation and Growth, edited by R. Rothwell and J. Bessant
Elsevier Science Publishers B.V., Amsterdam, 1987 — Printed in The Netherlands

CONSUMER INNOVATIVENESS: NOVELTY-SEEKING AND COGNITIVE STYLE

G R FOXALL

Department of Marketing, University of Strathclyde, 173 Cathedral Street, Glasgow G4 ORQ, Scotland (UK)

SUMMARY

Theories of new product development stress the importance of tailoring innovations to the requirements of their initial purchasers, consumer innovators. Attempts to identify cognitive and personality factors peculiar to innovators have been generally unsuccessful. This paper describes two empirical studies of consumer innovativeness which employed the Kirton Adaption-Innovation Inventory (KAI). The results indicate a positive and useful relationship between cognitive style/personality and consumer choice.

INTRODUCTION

The identification of the earliest adopters of new brands/products is of strategic importance to new product developers for two reasons. First, the prelaunch testing of innovations can be conducted with consumers who represent those who will initiate the diffusion process for the innovation. Secondly, marketing managers can tailor each element of the marketing mix to the requirements of this group and thereby accelerate early acceptance of the innovation. For such marketing planning, the most efficient means of identifying innovative consumers would be a psychometric test of innovativeness based on personality and/or cognitive preferences for novelty and creativity (refs.1-3).

However, in spite of progress in identifying the socio-economic characteristics of innovators, researchers have failed to isolate personality traits or cognitive processes which are consistently related to innovative buying. Although research involving dozens of psychometric tests and consumer purchase contexts has been conducted, the observed relationships between personality and consumer choice are extremely weak and do not form a basis for either academic theorisation or marketing strategy (ref.4). This research has produced such results for three reasons: first, it employed inappropriate tests of personality, developed and validated in clinical situations but having little or no relevance to economic decision-making; second, it was based on the assumption of a simple relationship between personality traits/types and customer behavior; third, the level of analysis chosen was inappropriate to investigating the psychological precursors of innovative behavior.

This paper describes two empirical investigations of consumer innovativeness which overcome these problems. The studies, of innovative brand choice for new foods, and innovative product choice for 'healthy foods', employed a measure of cognitive style, the Kirton Adaption-Innovation Inventory (ref.5). The results indicate that cognitive style is closely related to patterns of innovative purchasing of continuous and discontinuous new brands/products and to the volume of new brand/product purchasing engaged in by the earliest adopters. The results are significant for the development of a theoretical framework for the investigation of consumer innovativeness and for marketing strategy.

COGNITIVE STYLE AND THE CONTINUITY/DISCONTINUITY OF INNOVATIONS

Cognitive style refers to an individual's characteristic manner of processing information, especially in problem-solving and decision-making (ref.6). It describes the structure of thought rather than its content, the 'how' rather than the 'what' of thinking (ref.7), and is a hypothetical construct which accounts for observed regularities in behavior. Theories and measures of cognitive style rest on the assumption that such regularities in behavior can be attributed to regularities in information processing (ref.8). As posited intrapersonal structures which mediate stimulus and response to produce characteristic patterns of behavior, cognitive styles are closely related to traits of personality (refs.9,10).

Adaption-innovation theory states that individuals characteristically exhibit one or other of two cognitive styles, making decisions and solving problems either Adaptively or Innovatively. The intellectual processes of extreme Adaptors are confined by the frame of reference within which they perceive the activating problem to have arisen. Their definitions of both the problem and its frame of reference are inextricably linked and both remain intact while solutions are sought and the chosen solution implemented. Extreme Innovators, by contract, do not simply seek solutions to a known problem: they tend also to evaluate the frame of reference within which the problem has arisen and in the process of solving it they redefine and reconstitute both. Together, the Adaptive and Innovative cognitive styles described by Kirton and the KAI by which they are measured organize into a typology of higher order factors a whole series of traits, one of which, novelty-seeking, Hirschman (ref.11) has identified as centrally important. This typology offers, moreover, the opportunity of specifying the 'yet to be specified' components of Midgley and Dowling's (ref.12) concept of innate innovativeness.

The KAI comprises 32 five-point scales on which the respondent indicates the degree of ease or difficulty with which he feels he could consistently maintain specified Adaptive and Innovative behaviors over time. The behaviors include responding favorably to the stimulation of change and adherence by preference to the established rules. The theoretical range of individual scores is from 32 to 160. KAI scores for a large British general population sample (N=562) are normally distributed over an observed range of 45-146 with a mean of 94.99 (standard deviation, SD, 17.90) which approximates the theoretical mean of 96. KAI scores of females tend to be more Adaptive than those of males; in the UK general population sample, the 298 female respondents' KAI mean was 90.84 (SD 17.82) compared with make respondents' mean of 98.12 (SD 16.75). The difference is significant ($p<0.001$).

Cognitive style is conceptualised as a consistent mediator of behavior and correlates well with stable dimensions of personality: evidence is now available in the form of over sixty correlations, produced by more than twenty independent scholars in eight countries, involving over twenty different psychological tests (ref.13). 'The innovator tends to be more extravert, less dogmatic, more tolerant of ambiguity, more radical, more flexible, more creatively motivated, more creatively self-perceptive, more assertive, expedient, self-assured, undisciplined, independent and sensation-seeking than the adaptor; with more self-esteem, liable to risk-taking, needing (and liking) less structure, and is more spontaneous. The adaptor is more controlled, less stimulating, more steady, reliable, prudent and probably more often seen as right and dependable, better able to fit into teams, get on with authority, be sensitive to policy and mores; be more realistic, efficient and orderly. Neither type (extreme) is likely to be any more or less neurotic, more or less likely to reach high position (except in conditions unfavorable to type), be more or less intelligent, resourceful, original, creative and generally regarded in worldly terms as successful' (ref.14). Several KAI correlates are psychological and behavioral factors associated in the diffusion literature with the adoption of new ideas, practices and products. Relative to later adopters, early adopters (or 'consumer innovators') tend to be less dogmatic, more able to cope with abstractions, ambiguity and uncertainty, less fatalistic, more flexible, self-controlled and unsubjugated, and higher in achievement motivation and aspiration (ref.15). Innovators for consumer goods have frequently been described, on the basis of weak empirical evidence, as risk-takers, impulsive, active, dominant, inner-directed and self-reliant (ref.16). Goldsmith's (ref.17) empirical demonstration of the correlative relationships of Innovativeness, negatively with dogmatism, but positively with risk-taking and sensation-seeking is, therefore, especially pertinent. He also points to the tendency of Innovators in this sense to seek novelty, be inner-directed and intolerant of social norms. 'A consistent picture emerges of the intuitive-innovator who pays little attention to routine details, welcomes the new and different, and generates many novel ideas, contrasted with the equally-creative sensing-adaptor, who watches the details, works routinely and steadily, preferring standard solutions to problems' (ref.18).

The empirical research described below was stimulated by the possibility that consumer innovators (the earliest adopters of novel brands and products) would be cognitive Innovators (defined by Kirton's adaption-innovation theory). But there was no expectation of a simple, direct relationship for all brands and products investigated. So many investigations have discovered only slight correspondence between consumer innovativeness and personality, that the necessity of incorporating another variable was indicated in order to clarify the relationship between initial adoption and personality. Degree of continuity/discontinuity (the extent to which a new brand or product differs from established alternatives in the perception of prospective buyers) was hypothesized to be the required variable (refs.19,20). Continuous innovations are those which are least disruptive in their impact on consumer behavior compared with current alternatives (line extensions and product alterations, e.g fluoride toothpaste). Dynamically continuous innovations are those which have some disruptive influence on consumption behavior but which do not change established patterns fundamentally (e.g an electric toothbrush). Discontinuous innovations have considerably disruptive impact on consumption behavior or are associated with a basic change in lifestyle or consumption habits (items which are distinct in terms of taste, function or content: e.g video recorders or microwave ovens).

This reasoning led to the expectation that cognitive styles (Adaptive and Innovative) would be systematically related to patterns of innovative purchasing (i.e the number and type of innovation acquired by initial adopters) via the intervening variable, brand/product continuity/discontinuity.

METHOD

The empirical research comprised sequential surveys of (i) innovative consumers of 13 recently-launched brands of everyday food products (the Food Consumer Sample, or FCS), and (ii) innovative consumers of 24 classes of food product promoted as conducive to healthy-eating and recently launched through supermarkets (the Healthy-Eating Sample, HES).

FCS. A quota sample of 101 female supermarket shoppers, each of whom had purchased at least one of 13 innovative brands recently launched and available for approximately four months, completed the KAI schedule and provided additional information on their purchase behavior for the selected product classes. Each innovative brand was ranked by an ad hoc panel of female consumers according to its continuity/discontinuity with an existing member of its product class. The following classification resulted. Continuous brands: drinking yoghurt; cream and vegetable oil spread; cheese sauce granules, battered haddock steaks; and a mixed fruit drink. Dynamically-continuous brands: mixer sherry; reduced fat sausages; reduced fat cheddar cheese; fibre-rich cereal; and savory wheat cracker biscuits. Discontinuous brands: garlic and herb salad dressing; instant decaffeinated coffee; and a low calorie, sugar-free snack biscuit.

HES. A quota sample of 345 female supermarket shoppers completed the KAI and a questionnaire concerning their purchase of healthy foods. Healthy foods are versions of standard food products generally promoted as having a more beneficial/less deleterious effect upon consumers' health and wellbeing than the usual alternative. This product-level study was intended to show whether the KAI-based division of products into relatively continuous and discontinuous could be sustained at the product-class level. The study was limited to items available in supermarkets, whose purchase does not involve visiting specialist health food stores. During the two years preceding the research, TV programs and press reports, reinforced by official reports (refs.21,22) increased public awareness of the benefits of healthy eating. Most of the product classes had been introduced into supermarkets during this period; they and those which had been available there for some time previously (e.g wholewheat bread) had been strongly promoted in the months leading up to the investigation as supermarkets used their concentration on healthy-eating to attract interested customers from other outlets.

These hypotheses were tested:

H1: The KAI sample mean of (the female) purchasers of new brands/products is more Innovative than that of the female general population (FGP).

H2: There is no significant correlative relationship between consumers' KAI scores and the number of innovations purchased.

H3: The new brands investigated can be subdivided into groups in terms of their continuity-discontinuity with existing members of their respective product classes; and purchasers of continuous, dynamically continuous and discontinuous brands are distinguishable from each other by differences in their KAI scores.

H4: The healthy-eating products can be grouped similarly in terms of their continuity/discontinuity compared with established alternatives, while purchasers of relatively continuous and relatively discontinuous products can be distinguished from one another in terms of their KAI scores.
H5: More Adaptors than Innovators will be found among the purchasers of more continuous brands/products, while more Innovators than Adaptors will be found among the purchasers of more discontinuous brands/products.

RESULTS

FCS. Respondents' KAI scores are approximately normally distributed over a range of 54-145 which falls within but is rather more restricted than that of the general population. The KAI mean of the FCS is 95.06 (SD 15.82). For the purpose of the research respondents whose KAI score is lower than or equal to that of the FGP are referred to as Adaptors; those who score higher than the FGP as Innovators. The sample thus comprises 40 Adaptors and 61 Innovators; the observed range of new brand purchases is 1-8.

HES. Respondents' KAI scores are approximately normally-distributed over a range of 52-140, with a mean of 95.01 (SD 14.86). The sample contains 142 Adaptors and 203 Innovators; the observed range of products purchased is 0-19; four respondents reported having bought none of the items under investigation.

Hypotheses 1 and 2 are accepted. The mean of each of the samples is significantly more Innovative than that of the UK female general population: FCS mean = 95.06 (SD 15.82), z = 2.272, $p<.025$; HES mean = 95.01 (SD 14.86), z = 3.195, $p<.001$. In the case of the FCS, there is a weak positive but non-significant correlation between the number of new brands purchased and KAI score: the Pearson product moment correlation coefficient, r = .091 (ns). Similarly, in the case of the HES, the Pearson product moment correlation coefficient, r = .041 (ns).

The relationship between KAI score and number of innovations purchased is not simple, however. While the KAI mean of FCS purchasers of 4/5 new brands is significantly different from that of the FGP, the mean score of neither the purchasers of 1-3 brands nor purchasers of 6-8 brands differs significantly from that of the FGP. Nor does the mean score of any of these subgroups of purchasers differs significantly from that of any other. Further, neither the KAI mean of HES purchasers of up to two new products nor that of purchasers of 16-19 differs significantly from that of the FGP; the means of purchasers of other quantities of new products are, however, significantly higher than that of the FGP. The KAI mean of purchasers of up to two new products is, like that of purchasers of more than 16, distinctly Adaptive.

The means of purchasers of intermediate numbers of products (3-5, 6-8, 9-11, 12-15) all significantly exceed that of the FGP. Moreover, while the KAI means of purchasers of up to two new products and of 16-19 new products do not differ significantly from one another, each differs significantly from those of purchasers of every other quantity.

Hypothesis 3 is accepted. The means of purchasers of the panel-determined discontinuous and dynamically-continuous brands (taken separately or combined) differ significantly from the mean of the FGP; that of purchasers of the continuous brands does not. None of the means of the subsamples based on brand continuity/discontinuity differed significantly from another. Further, analysis of the data revealed that the 13 brands could also be placed in three groups based on the patterns of KAI means of the purchasers and non-purchasers of each. The groups so-constructed differed only marginally from those suggested by the panel. The general pattern was of the mean score of consumers of more discontinuous brands being Innovative while that of consumers being attracted by continuous brands did not differ significantly from that of the FGP. GROUP 1: The KAI mean of purchasers of each brand in this category was (a) significantly more Innovative than that of the FGP (though that of non-purchasers of the brand who had nevertheless purchased from the product class was not); and (b) significantly more Innovative than that these non-purchasers of the brand. The three brands included in this group were identical to those classified as discontinuous by the panel. GROUP 2: The KAI means of purchasers of each of the brands in this category were in every case more Innovative than that of non-purchasers of the brand (who had nevertheless bought within the relevant product class). None of the differences in means was significant, however. There was a mixed pattern of relationships between the KAI means of buyers and non-buyers of each brand compared with that of the FGP. Group 2 comprised six brands, four of which had been identified by the panel as dynamically-continuous. The Group 2 brands were of mixer sherry, fibre-rich cereal, reduced fat fork sausages, battered haddock steaks, drinking yoghurt, and savory wheat crackers. GROUP 3: The KAI means of purchasers of each of the brands in this category fell below the sample mean; none was significantly different from that of the FGP. The KAI mean of non-purchasers of each of the four brands (who had, however, purchased competing brands) was in each case significantly more Innovative than that of the FGP. For each of the four brands, the mean of non-purchasers just exceeded the corresponding mean of purchasers of the brand. The latter difference was significant in only one case, however, that of the cream and vegetable oil spread, where the purchaser mean was distinctly Adaptive. Three of the four brands which comprised Group 3 were also classified as continuous by the consumer panel. The Group 3 brands were of reduced fat cheddar cheese, cream and vegetable oil spread, mixed fruit drink, and instant cheese sauce granules.

Overall, for 10 of the 13 brands investigated the panel's judgement of continuity-discontinuity was confirmed by the KAI analysis of brands. Furthermore, comparisons of the means of these three groups with each other and with the mean of the FGP shows a pattern similar to that produced by the panel-determined classification.

Comparisons of the means of consumers who purchased exclusively within each of the groups determined by inspection of the data shows that all differ significantly from that of the FGP. However, while the mean of purchasers of Group 1-only and that of purchasers of Group 2-only brands are highly Innovative, that of purchasers of Group 3-only brands is distinctly Adaptive. Further, the mean of Group 3-only purchasers is significantly more Adaptive than that of purchasers of Group 1-only purchasers and that of Group 2-only purchasers. That significant differences are found among such small sub-samples attests to the robustness of the results and the underlying relationships they reveal.

Three product groupings also emerge from analysis of the data for the HES; less clear-cut than in the case of the innovative brands, they are consistent with the expected theoretical pattern. GROUP A: The KAI mean of purchasers of each of the products in this category was significantly more Innovative than that of non-buyers. For both of the products in Group A, the KAI mean of buyers was significantly greater than that of the FGP. Group A contained two of the 24 products investigated: the wholewheat snack biscuit and wholewheat pasta. GROUP B: The KAI mean of purchasers of each of the products in this group is more Innovative than that of non-buyers but the difference in no case reaches the conventional level of significance. For Group B products, as for those of Group A, the means of buyers always exceed 95, the sample mean, while those of non-buyers never reach this point. In all but two instances, the KAI means of both buyers and non-buyers of Group B products are significantly more Innovative than that of the FGP. Group B comprised 14 products: wholewheat bread; low calorie soft drinks; high-fibre cereal; semi-skimmed milk; low fat, low sugar yoghurt; sunflower margarine; decaffeinated coffee; sugar-reduced jam; tofu; calorie-reduced salad dressing; sugar and starch reduced baked beans; saccharine-free sweetener; 300-calorie meal; and tuna canned in brine. GROUP C: The KAI mean of buyers of each product in this group is more Adaptive than that of non-buyers but none of the differences in means reaches statistical significance. For every Group C product, non-buyers' means exceed 95 but this is nowhere true of buyers. For 5 of the 8 products, buyers' KAI means are not significantly different from that of the FGP. The eight products in Group C are calorie reduced soup; low fat sausages; sunflower cooking oil; skimmed milk; low fat cheese; wholewheat potato chips; low sodium salt; and low fat cream.

While the KAI means of purchasers of all three groups of products significantly exceed that of the FGP, they decrease directly from Group A, through Group B, to Group C. None of the groups' means differs significantly from another's, though the difference between the means of Groups A and C is large and in the hypothesised direction. Groups of exclusive purchasers of each of the categories of product could not be established since there was considerable overlap between purchases.

Hypothesis 5 is accepted for brands but not products. For the FCS, Innovative purchasers per brand outnumber Adaptors on average for Groups 1 and 2, while Adaptors outnumber Innovators on average in the case of Group 3 brands (groups determined on inspection of data). Grouping the data to take account of the lack of any statistical difference between the KAI means of Groups 1 and 2, to give a 2 x 2 matrix, indicates that the difference in numbers of Adaptors and Innovators buying relatively continuous and discontinuous brands is significant. Moreover, the numbers of Adaptors and Innovators buying each type of brand expressed as percentages of the numbers of Adaptors and Innovators respectively in the sample differ significantly. For the HES the differences in the numbers of Adaptors and Innovators purchasing relatively continuous and discontinuous products is not statistically significant, though the ratio of Adaptors to Innovators declines as expected from Group A (approximately 1:2) through Group B (approximately 2:3) to Group C (approximately 3:4). Nor is the difference between the average numbers of Adaptors and Innovators per brand in each group expressed as a proportion of their respective components of the HES. (Full details of the results, which are only summarised here, are in ref.20).

DISCUSSION

Both Adaptors and Innovators. Consumer innovators taken as a whole show a tendency to be Innovative; on average, they are significantly more so than members of the female general population. But the straightforward correlation of the KAI scores of the members of the FCS and HES samples respectively with the number of brands or products they have purchased indicates a level of association which is no more than typical of the dozens of similar findings of positive but inauspicious relationships between personality and buyer behavior. The results confirm that any simplistic search for a personality trait (or group of traits) or type which is ubiquitously possessed in a direct and straightforward manner by consumer innovators but lacking in later adopters is likely to fail.

However, the research has demonstrated that both Adaptive and Innovative cognitive styles are found among the first adopters of the food brands and products examined: knowledge of both of these bipolar styles, and of the distinct personality and behavioral characteristics associated with each, is necessary to a full understanding of the role of cognitive and personal factors in consumer innovativeness. Both Adaptors and Innovators are attracted to 'newness', though this has a different meaning for each. For the Adaptor, newness involves insight within an existing frame of reference, exploration within that paradigm, perhaps even its extension. For the Innovator, novelty implies a break in the paradigm or perhaps a new synthesis of two existing paradigms which effectively breaks both. Given this, the usual search for a unipolar mental or personal ability or component which explains innovative behavior in global terms is naive as well as inadequate.

Continuity/Discontinuity. It is only when brands/products are subdivided according to their continuity/discontinuity (initially on the basis of consumer perceptions and subsequently by analysis of the data) that the robust relationship between cognitive style and innovativeness becomes clear. The KAI mean of purchasers of relatively discontinuous innovations is highly Innovative; more Innovators than Adaptors purchase such items. The KAI mean of purchasers of relatively continuous innovations does not differ from that of the FGP; more Adaptors than Innovators purchase such items. These tendencies are seen more clearly in the case of the brands studied, where the means of subsamples of respondents who purchased exclusively either discontinuous or dynamically-continuous or continuous innovations clearly reinforce the pattern. Moreover, purchasers of continuous innovations only differed in terms of mean KAI scores from purchasers who confined their buying to either of the other groups of brands. While the KAI means of all three groups established on the basis of exclusive purchasing differed significantly from that of the FGP, those of the discontinuous-only and dynamically-continuous-only brand purchasers were significantly more Innovative than that of the female general population, that of the continuous-only brand purchasers was significantly more Adaptive.

Qualitative Differences between Adaptors and Innovators. While both Adaptors and Innovators provided large proportions of the respondents who reported having purchased each brand/product, irrespective of its degree of continuity-discontinuity, they did so in varying proportions according to this variable. There is a tendency for Innovators to constitute a higher proportion of buyers of the discontinuous brands/products and for Adaptors to constitute a higher proportion of buyers of continuous brands/products. In addition, a higher proportion of the Innovators in each sample was attracted to the more discontinuous brands/products, while a higher proportion of the Adaptors in the sample was attracted to the continuous innovations. This coexistence of Innovators and Adaptors among the earliest adopters of innovations accounts for the generally poor correlations and contradictory evidence reported by studies of personality links with consumer innovativeness. Many traits weakly associated in research with consumer innovativeness - e.g risk-taking and inner-directedness - appear to be characteristic of only one subgroup of consumer innovators, the Innovators.

Quantitative Differences between Adaptors and Innovators. Adaptors and Innovators differ quantitatively as well as qualitatively. For both brands and products, the KAI mean of purchasers of the fewest innovations was Adaptive and means of purchasers of intermediate numbers of innovations were distinctly Innovative. However, the mean of purchasers of the largest numbers of brands or products were again Adaptive. Though these high-volume purchasers constitute small subsamples, their KAI means are in both cases lower than those of the purchasers of the smallest number of items, almost identical to the mean of the FGP in the case of brands and considerably below it in the case of products. This pattern reveals the relevance to the proper understanding of consumer innovativeness of both cognitive styles under investigation. Innovators are generally willing to experiment and, in the context of consumer choice, this takes the form of frequently trying new brands and products, possibly on impulse.

It is the Adaptors who are more usually the systematic seekers-out of a large number of new products. In the case of healthy-eating, the logic of this is easily demonstrable: Adaptors are wary of dramatic change and avoid the experimentation and risk-taking which impulsive innovative buying involves; but, once 'converted' to the cause of healthy-eating (perhaps as part of a more healthy style of living in general), they systematically, carefully and comprehensively substitute new versions of products for their initially more familiar, tried and known, but now inappropriate correspondents. The same sound and systematic approach to all their activities which initially predisposes them to avoid highly innovative products, is now manifest in their assiduous search within the new frame of reference for the full range of appropriate products.

The inevitable conclusion is that consumer innovativeness may be mediated by one of two cognitive styles, each associated with a distinct pattern of personal and behavioral traits and expressed in distinctive patterns of brand/product quality and quantity. Cognitive style should, therefore, be incorporated into consumer theory as a measurable explicative variable, while marketing managers should recognize the need to create and implement two innovator-directed launch marketing mixes.

REFERENCES

1. M J Baker, Market Development, Penguin Books, Harmondsworth, 1983.
2. G R Foxall, Corporate Innovation: Marketing and Strategy, Croom Helm, London; St.Martin's Press, New York, 1984.
3. D F Midgley, Innovation and New Product Marketing, Croom Helm, London, 1977.
4. H H Kassarjian and M J Sheffet, Personality and consumer behavior: An update, in H H Kassarjian and T S Robertson (eds) Perspectives in Consumer Behavior, Scott Foresman & Co, Glenview, IL, 160-80.
5. M J Kirton, Adaptors and Innovators: A description and measure, J. Appl. Psychol, 61 (1976) 622-629.
6. F P McKenna, Measures of field dependence: Cognitive style or cognitive ability? J. Person. & Soc. Psychol. 47(3) (1984) 593-603.
7. K M Goldstein and S Blackman, S. Cognitive Style: Five approaches and relevant research, Wiley, New York, 1978
8. K M Goldstein and S Blackman, Cognitive style, in: F Fransella (Ed), Personality: Theory, Measurement and Research, Methuen, London, 1981.
9. J P Guilford, Cognitive styles: What are they? Educ & Psychol. Measure, 40 (1980) 715-35.
10. N Kogan, Creativity and cognitive style: A life-span perspective, in: P B Baltes and K W Schaie (eds), Life-span Developmental Psychology: Personality and Socialization, Academic Press, New York, 1973.
11. E C Hirschman, Innovativeness, novelty seeking, and consumer creativity, J. Cons. Res, 7 (1980) 283-95.
12. D F Midgley and G R Dowling, Innovativeness: The concept and its measurement, J. Cons. Res, 4(1978) 229-242.
13. M J Kirton, Adaptors and innovators: Problem-solvers in organizations, in: K. Gronhaug and G Kaufman (Eds), Innovation: A Crossdisciplinary Perspective, Wiley, New York, 1987a, in press.
14. M J Kirton, Adaptors and innovators: Cognitive style and personality, in: S G Isaksen (ed), Frontiers of Creativity, Brearly Ltd, Buffalo, NY, 1987b, in press.
15. E M Rogers, Diffusion of Innovations, 3rd ed, Free Press, New York, 1983.
16. D F Midgley, op. cit.
17. R E Goldsmith, Personality characteristics associated with adaption-innovation, J. Psychol, 117 (1984) 159-65.
18. R E Goldsmith, Personality and adaptive-innovative problem solving, J. Soc. Behav & Person, 1 (1985) 95-106.
19. T S Robertson, The process of innovation and the diffusion of innovation, J. Market, 31, 1967) 14-19.
20. G R Foxall, Consumer innovativeness: Novelty-seeking, creativity and cognitive style, in: E C Hirschman (ed) Research in Consumer Behaviour, 3, Greenwich, CT: JAI Press (in press).
21. COMA (Committee on Medical Aspects of Food Policy), Diet and Cardiovascular Disease, Department of Health and Social Security Report No.28, HMSO, London, 1984.
22. NACHE (National Advisory Committee on Health Education), A discussion paper on proposals for National guidelines for Health Education in Britain, Health Education Council, London, 1983.

Innovation: Adaptation and Growth, edited by R. Rothwell and J. Bessant
Elsevier Science Publishers B.V., Amsterdam, 1987 — Printed in The Netherlands

STRATEGIC IMPLICATIONS OF USER-INITIATED INNOVATION

G R Foxall
Department of Marketing, University of Strathclyde, 173 Cathedral Street, Glasgow G4 0RQ, Scotland (UK).

SUMMARY

User-initiated innovation occurs when a firm exploits a novel device which it has invented, either by investing in its internal application as a process innovation (benefiting from the enhancement of output), or by its marketing and diffusion as well (gaining income through fees, royalties or profits). Case studies are described which identify several modes of user-initiated innovation and users' roles in them. Three possible explanations of the functional locus of innovation are critically examined: the appropriability of benefit, appropriability regimes, and transaction costs hypotheses. Elements of all three are required in order to understand user initiated innovation.

THE LOCUS OF INNOVATION

The usual understanding of the roles of manufacturers and users in industrial innovation attributes the initiative for new product development to the former. According to the manufacturer-active paradigm or MAP (ref.1), the manufacturer's role is to survey a group of customers, obtain ideas for new products, analyze the market data and develop and market a responsive product. In contrast to this dominant view, there is evidence that, in some circumstances (e.g when suppliers cannot provide rapid solutions to technical problems) the development of a new industrial process is open to user- rather than manufacturer-initiation. The customer-active paradigm (CAP) describes this situation in which the user/customer generates the idea for the innovative process, perhaps going on to design, produce, apply and test an in-house prototype before approaching a manufacturer who, if he perceives a market opportunity, produces and sells it.

The CAP extends the perspective often adopted in industrial innovation studies by drawing attention to a neglected aspect of the role of users in NPD, but, even here, the entrepreneur, alert to the chance of personal gain and maximizing the economic benefits of marketing the product innovation, is the manufacturer. The user is an inventor and process innovator, but not a product innovator who gains from the diffusion of the innovation. Whilst he may receive token royalties from the commercialization of his invention, his interest lies in securing a continuous and trustworthy source of supply.

A third paradigm acknowledges the entrepreneurial involvement of some industrial consumers in the commercial exploitation of product innovations they have previously devised and applied internally as process innovations (ref.2). It describes a situation in which the user assumes some or all of the entrepreneurial role ascribed by both MAP and CAP to the producer. When the user-initiator engages in design work or prototype building and testing or development with the intentions of enhancing the attractiveness of his invention to other prospective users, and of increasing his gains as a result, he evinces entrepreneurial alertness with respect to the marketing of his invention as a product. The user-initiator may go further: forecasting demand for the product innovation, protecting the intellectual property rights in his invention, and strengthening his hand in negotiations with potential licensees. Thereby, the user-initiator actively engages in user-initiated product innovation, and behaves in ways that transcend the CAP.

Empirical research (ref.3) comprising detailed case studies of user involvement in industrial NPD confirms this paradigm. This paper is based on two such cases - (i) the work of the Business Development Group at the Warton Unit of British Aerospace (BAe); and (ii) the development of control software for flexible manufacturing systems (FMS) at BAe in joint venture with the UK General Electric Company (GEC) - which demonstrate a range of user-manufacturer relationships in industrial innovation, extending from that in which the manufacturer has exclusive control of the NPD sequence to that in which the user assumes such control. Between these extremes are situations in which the user-initiator exploits his self-devised process innovation without reference to other firms and those in which he pursues quasi-integrated vertical relationships (ref.4) to exploit jointly the benefits of marketing the innovation as a product. Five patterns of industrial innovation are defined and illustrated below in the context of the case histories.

USER-INITIATED INNOVATION

BAe's Warton Unit, in Lancashire, England, which is responsible for military aircraft, makes strenuous efforts to maintain its position in advanced manufacturing techniques (AMT). When they encounter novel problems, BAe engineers often enter into collaborative ventures with external organisations to design, develop and produce the required device. Such collaboration has resulted on occasion in the intellectual property rights to the jointly-produced innovations being secured by the external partner: Warton has obtained supplies of the device, but the co-developers have received the commercial benefit of its being generally marketed. As a result Warton established a Business Development Group (BDG) to ensure that it benefited directly from commercial exploitation of its inventions (ref.5).

The BDG identifies internally generated and applied process innovations with potential for commercialization, patenting and/or copyrighting them before seeking firms to co-develop, manufacture and market the resulting products. When it cannot quickly license-out, the BDG assesses potential demand in conjunction with the personnel responsible for the invention and its process innovation. Information is also acquired on unique qualities and user benefits of the device, increasing the probability of concluding a profitable licensing agreement with an external manufacturer. On the basis of this knowledge, the BDG, wherever possible, negotiates the licensing of the innovation and subsequently monitors the agreement.

The manufacturer-user interactions found are described and exemplified below.

Manufacturer-Initiated Innovation (MII) The manufacturer is responsible for the management of the NPD process in its entirety, from the generation and screening of ideas through concept testing and evaluation to product development, market testing and commercialization. This procedure (which corresponds to the MAP) is the pattern of industrial NPD traditionally taken for granted by most marketing authors.

User-Initiated Process Innovation (UII1). The user invents and develops a novel device for internal process innovation which has no external significance (except as it is embodied in that firm's output). E.g Warton Unit produced a plastic electric wiring loom pin with the intention of marketing it more generally. No external market was found for this device which is applied as an internal process innovation.

Passive User-Initiated Product Innovation (UII2). The user-initiator of an internally-developed and implemented process innovation approaches a manufacturer with an idea, design, or tried and tested prototype requesting him to produce it. The manufacturer responds only if the demand from the user-initiator is large or he expects to exploit a large general market for the device. The user may receive royalties or fees, but the manufacturer acts entrepreneurially in commercializing the invention. As in the CAP, the user is passive where product innovation is concerned, a customer rather than a marketer, interested in the continuity of his in-house use of the invention rather than its external diffusion. E.g An aluminium cutting tool was developed by BAe with an outside collaborator, who went on to obtain for himself the intellectual property it represented and commercialized it generally, leaving BAe with no more than its original process innovation.

Active User-Initiated Product Innovation (UII3). The user acts entrepreneurially in the commercial exploitation of his process innovation, actively seeing to gain from its diffusion as a product. The user does not himself produce or market the item but develops as strong a negotiating position as possible vis-a-vis the external manufacturer with whom he becomes quasi-vertically integrated, a relationship based on and sustained by his unique access to technical and market knowledge. His involvement may include patenting, demand analysis, and monitoring contracts. The emphasis is upon gaining from the external exploitation of the process innovation as a widely-diffused product. E.g The self-normalising test probe, invented at Warton, identifies structural flaws in materials such as carbon fibre composites. It is superior to existing test probes in that it remains normal at all times to the test piece whose contours it follows without prior 'knowledge' of its shape. BAe engineers suggested such a device to their supplier of ultrasonic test equipment; the development of the self-normalising probe by this supplier was financed by BAe and the companies patented the device jointly. The supplier manufactures and markets the probe and, under a formal licensing agreement, pays BAe a percentage of the resulting sales revenue.

Vertically-Integrated User-Initiated Product Innovation (UII4). The user-initiator is responsible for all stages of the industrial NPD sequence. Having developed a process innovation, he uses it in his own productive processes and goes on to manufacture and market it. The entrepreneurial involvement of the user is complete: he has not only been a consumer of his own technology but assumes control of the whole process by means of which that technology is marketed. E.g BAe attempted to purchase control software for flexible manufacturing systems from external sources but none had then developed the necessary technology. It was proposed that the Unit would produce the required software to control its own manufacturing processes but financial considerations compelled the unit to develop a system which could be profitably commercialized. The unit entered into a joint development with GEC Electrical Products Ltd, which had also been considering the development of similar technology. Initial sales of the developed technology were within the two companies but external organisations expressed interest in purchasing the system even before it was available. Following the successful joint venture program, the companies decided to market various products and services independently rather than by establishing a jointly-owned sales subsidiary (refs.6,7).

EXPLAINING USER-INITIATED INNOVATION

It is necessary to explain why the functional locus of innovation often lies with users rather than with manufacturers and why users committed to benefiting from the marketing of their process innovations choose specific management structures (quasi- and full-vertical integration) to accomplish this. Three explanations are relevant.

Appropriability of Benefit Hypothesis. Von Hippel (ref.8) predicts the functional locus of innovation in terms of the differential ability of manufacturers and users to benefit from the sales of output-embodied innovation knowledge. There is little chance of innovators benefiting from non-embodied innovation knowledge (its transfer via sale or licensing as information rather than in the form of products). Patents afford little protection to the innovator: obtaining redress for infringement is costly and slow, and competitors can 'invent around' patents. Nor do trade secrets usually provide protection: other firms may independently discover the technology, obtain it by accidental disclosure of 'reverse engineering'.

The alternative is to benefit from output-embodied knowledge (the in-house application by an innovator of his new process and its subsequent enhancement of the value of his output). The innovator's capacity to do so depends upon his ability to set up a quasi-monopoly on the basis of his innovation, either at the industry-level so that competitors share his knowledge but he (presumably as possessor of dominant market share) benefits most; or at the firm-level so that the innovator alone benefits. As in the case of non-embodied knowledge, patenting and know-how afford little protection and the chief source of benefit sustained by a firm-level quasi-monopoly is the time required by competitors to respond imitatively; such 'response time' provides the innovator with a temporary monopoly. Manufacturers can establish quasi-monopolies only by means of secretable innovations which they employ in the building of their machinery: the amount of such secret knowledge is directly proportional to the experience they gain in producing machines, i.e to the number of machines produced. Therefore, only when manufacturers of process machinery can expect to product and sell innovations on a scale much greater than the demand represented by a single large user is the locus of innovation predicted to lie with the manufacturer. Otherwise, it is predicted to lie with the user.

Regime of Appropriability Hypothesis. Teece (ref.9) sets out to explain why the economic returns of technological innovation may accrue not to the innovator but to an imitator in terms of the relative strength of the 'regime of appropriability' (i) surrounding the innovator and potential imitators. Also, the stage of development of industry dominant designs (ii) and the availability of complementary assets (iii) determine whether the innovator benefits. (i) Like von Hippel, Teece argues that patents and trade secrets protection rarely provide effective protection: in some industries technological knowledge is easily codified and thus open to articulation and transmission. (ii) With Abernathy and Utterback (ref.10) he argues that the development of some innovations has two phases, the preparadigmatic which obtains before a dominant design emerges, and the paradigmatic which follows the general acceptance of such a design. In the first, the innovator may lose out to an imitator who develops the innovator's idea to a generally acceptable stage and thereby establishes the paradigm to the disadvantage of the originator. (iii) Of particular relevance to this paper is the observation that the profitable commercialization of a core innovation technology often depends upon the use of complementary assets such as marketing and after-sales service. The more specialized these are to the core technology, the greater will be the need for the innovator to obtain and protect them in order to gain from the diffusion of his innovation.

Tight appropriability regimes are unusual since patents, copyrights or trade secrets rarely prevent imitation, though where they do obtain, the innovator gains significantly. If specialized or co-specialized assets (the latter involve bilateral dependence between say a manufacturer and distributor) are involved, the innovator should integrate manufacture and/or distribution in order to protect his investment against a partner's withdrawal. Weak appropriability is characterised at the preparadigmatic stage by there being few complementary assets but these become increasingly important as the paradigm design becomes established. Once the design is generally accepted and the core technology is imitable, control of such assets is crucial to the strategic exploitation of the innovation.

Transactions Costs Hypothesis. Transactions cost analysis, which is concerned to ascertain the governance structures appropriate to productive tasks, rests upon two behavioral assumptions: (i) the bounded rationality of decision-makers' cognitive processes and consequent actions, and (ii) the opportunism to which market arrangements may be subjected by one of the transacting parties. The impetus to internalize production/distribution decisions and actions, i.e to substitute hierarchical, intra-firm control based on sequential, adaptive decision-making for market exchanges based on contingent claims contractual bargaining as the means of governing transactions is predicted to increase directly with increases in transactions costs deriving from (a) uncertainty, (b) the frequency of transactions, and (c) the transaction-specificity of assets. Once the costs associated with these factors reach a certain level, vertical integration provides a superior governance structure to market transacting by its provision of incentives and controls, and its facilitation of organisational information processing (refs.11,12).

Holding constant the effects of uncertainty and transactions frequency, it is predicted that, where assets lack specificity to either transacting party, classic market contracting will obtain; at semi-specific levels of asset specificity, obligational market contracting will occur; and where assets are highly specific, markets will be replaced by internal organisation. Similarly, holding constant the effects of asset-specificity and transactions frequency, increasing uncertainty will lead to predictions of a similar progression of governance structures from the simple to the complex (ref.13).

Transactions cost reasoning applied to industrial innovation must acknowledge two variables (ref.14): (i) the extent of the complexity involved in different technological innovations which is predicted to increase directly with the complexity of the appropriate governance structure for its development and production, and (ii) the possibility of intermediate forms of integration between users and manufacturers in the development of particular novel technologies. For market contracting to occur the user must be able to specify in detail the technology he requires, and the number of suppliers of such technology must be relatively large. Contingent contracting is the probable governance structure in this case because the costs involved in tendering contracts are lower than the potential costs of administering integrated organisation. But, when the development of technology is exploratory, few suppliers are able to develop and market the technology, and the advantages of being the first mover accentuate the bargaining problems inherent in small numbers by making opportunism more probable, more formally-integrated R & D, development/production/marketing is probable. Typically it takes the form of quasi-vertical integration which assures the supplier of sales and that his transfer of information to the user will not enable the user to produce his own technology or favor other suppliers. The more discontinuous the innovation under development, the greater the tendency towards vertical integration. Hence competitive bidding is the most probable governance structure when precise specification of the required performance and features of the technology is possible; when such specificity is less feasible, quasi-integrated contracting is probable; and when discontinuous technology which serves a particular user's needs is under development, the appropriate governance structure is vertical integration.

In summary, the hypotheses suggest the following.

Appropriability of Benefit: The innovator will benefit from his innovation knowledge when (a) that knowledge is output-embodied, and (b) he can achieve a quasi-monopoly on the basis of sufficient response time. It is highly improbable that he will benefit from non-embodied innovation knowledge since he cannot long protect such intelligence.

Appropriability Regimes: The innovator in a weak appropriability regime will obtain commercial advantage from his innovation when he has access to specialised complementary assets.

Transaction Costs: The adoption of complex governance structures will be (a) directly proportional to the discontinuity of the innovation under development and thus to the technical and market uncertainty surrounding its commercialization, and (b) inversely proportional to the numbers of producers and potential buyers.

DISCUSSION

The following discussion examines these arguments as explicators of user-initiated innovation.

Appropriability of Benefit. This hypothesis explains the functional locus of innovation which occurs within a vertically integrated structure, i.e when either a manufacturer or a user has full control over the NPD procedure. In MII, the user presumably cannot capture benefit from marketing the product, in UII1 there is no mechanism in the face of lack of market demand for the user to diffuse his invention commercially, and in UII4, the user appropriates benefit by himself becoming the product's supplier.

In UII2, exemplified by the aluminium cutting tool, the hypothesis correctly places the functional locus of innovation with the manufacturer since the user-initiator has no means of protecting his intellectual property. But the hypothesis does not explain active user-initiated product innovation: it follows from its reasoning that UII3 is highly unlikely since the innovator cannot protect the non-embodied innovation knowledge upon which it depends. In UII3, the possibility of the user-innovator effectively patenting his innovation or protecting it by concealing know-how is minimal; he must reveal technical and market knowledge to the manufacturer in order to conclude a mutually-acceptable licensing agreement; if the manufacturer has been a co-developer and co-patentee of the device, he has much of this knowledge anyway. Such active user-initiated product innovation, in which the user-initiator benefits satisfactorily from the general diffusion and commercial exploitation of his process innovation in spite of his apparent inability to form a firm-level quasi-monopoly, nevertheless occurs with some frequency. The appropriability of benefit hypothesis predicts that capture of non-embodied innovation knowledge is improbable rather than impossible but offers no assistance in predicting or understanding why it exists where it does.

Regimes of Appropriability. Nor does this hypothesis predict active user-initiated product innovation which requires that the new technology must be revealed in some degree to possible co-partners who might act opportunistically by becoming imitators. Further, user-initiated core technology knowledge is often codified and communicable. As a hypothesis suggests, integrated governance structures are sought more when the user-initiator is establishing a dominant design as in the case of the FMS control software project, than when it is accommodating its technology incrementally within a known design framework based on known user requirements, as with the aluminium cutter and the self-normalising test probe. The FMS software development project included the emergence of a dominant design for at least a subset of the industry. firms were attempting to buy the product even prior to its development, suggesting some willingness to accept it as an industry standard. Moreover, implementation and application knowledge constituted an important complementary asset. The UII2 and UII3 examples, by contrast, are of completed products whose imitability is moderately high and where complementary assets are more obvious and more easily appropriated by an opportunistic co-partner.

Transactions Costs. A broader explanation of the observed pattern of user-initiated innovation emerges from the transactions cost approach. Generally, patterns of integrated user-initiated process/product innovation (UII1, UII4) are associated with the highest levels of uncertainty and asset-specificity. The discontinuity of the innovations and the small numbers of buyers/sellers either preclude general marketing while permitting the user-initiator to benefit from enhanced output-embodied knowledge via the internal application of his invention (UII1) or increase the likelihood of post-contractual opportunism if classic market trading is attempted (UII4). In both cases, the observed governance structure is the integration which the transactions cost approach predicts.

Intermediate governance structures (the quasi-integration possibly attempted in UII2, but achieved by the more active user (UII3)) are associated with intermediate levels of uncertainty and asset specificity. In UII2, commercial initiative passes to the manufacturer rather than the user-initiator who has only his patents or know-how, notoriously poor guardians, to protect his interests and is exposed to the danger of providing the manufacturer with enough technical and perhaps market information to render the user superfluous. In the case of the aluminium cutting tool, the user lost the initiative, whilst the manufacturer's propensity to act opportunistically was high: the dynamically-continuous technology involved, coupled with moderate asset specificity and a large number of potential buyers made this likely.

The quasi-vertical integration found in UII3 is appropriate to the moderate-to-high asset specificity and technological uncertainty of the innovations represented. Potential manufacturers who might compete with BAe are, in the short term, few in number and there are non-trivial first mover advantages. The technology is relatively discontinuous and entails high uncertainty, and there is a considerable possibility of opportunism on the part of licensees. The user-initiator, who makes information available to manufacturers, must, in order to develop a successful quasi-vertically integrated relationship with a manufacturer, make full use of the information about the implementation and use of the process innovation if he is to benefit from its marketing. When the product is launched by the manufacturer, the user-initiator understands better than any other company can the problems and potential of investing in the technology, for he alone has applied it productively. Even a firm which makes similar products will not necessarily have access to such detailed information about the practical application of the user-developed innovation. This knowledge is crucial not only to the firm's capacity to exploit its technology itself but to its effective diffusion. It should be regarded not as trade secrets - since it must in course of time be revealed to an external firm, but as a <u>specialised</u> complementary asset, <u>used</u> by the manufacturer to market the new product but <u>owned</u> by the user: the essence of effective quasi-vertical integration (ref.15). The complementary assets and response time provided by such knowledge underpin the manufacturer's marketing program and are the user's only means of preventing the manufacturer's opportunism.

Quasi-vertical integration is attractive to the manufacturer when it provides a means of obtaining the user-initiator's knowledge on the match between the market and the new technology and forms the basis of response time advantages. The manufacturer can secure information to guide his own developmental, manufacturing and marketing operations from the user-initiator alone. The more discontinuous the technology, the less precisely will market needs be specifiable and the greater the importance of the user-initiator who represents the market place (ref.16). Thus the manufacturer is bound by his contract with the user-initiator, unable to act opportunistically, and the user gains from the marketing of his innovation. Hence the different experiences of BAe in the cases of the self-normalising test probe and the aluminium cutting tool. In the former case, BAe could control the revelation of information unavailable from any other source on the operation of the innovation: their consequent ability to create a firm-level quasi-monopoly based on the manufacturer's parallel need of a response time-based quasi-monopoly explains the mutually-satisfactory interaction. In the latter case, BAe had not developed the product to the extent of fully implementing it and thus lacked the knowledge required to forestall opportunism.

In UII4, as befits a situation in which there are few suppliers and, initially, few customers, the governance structure is vertical integration. Uncertainty is high since the complex technology under development is discontinuous: its nature and possible applications cannot be specified in detail ex ante. There is very high asset specificity especially with respect to the human resources which will provide installation and after-sales service, and there are enormous first mover advantages. Thus there is scope for postcontractual opportunism deriving principally from the absence of alternative suppliers. The transactions costs hypothesis is borne out by the joint venture between the two firms at broadly similar stages of technological development for control software. The partners thereby retained complete vertical control of the enterprise and were able to carefully monitor and protect it. The joint venture provided the possibility of sharing development costs but also of establishing more quickly and surely the technical guideposts to evaluate early R&D and design work. Finally, the firms chose horizontal disintegration once the technology had been created and the specific roles of the two companies in marketing versions of the technology became apparent. Once the commercial benefits of the project became realizable, they explicitly rejected the establishment of a jointly-owned marketing organisation and so further internalised control of production and marketing.

CONCLUSION

The research discussed above extends the MAP/CAP dichotomy and makes necessary a thorough re-examination of the roles traditionally ascribed to users and manufacturers in industrial innovation. Even in the 'non-entrepreneurial' modes of user-initiated innovation there is evidence that the intent of the user may be to benefit commercially from the marketing of his invention as well as from its internal application. This is at odds with the user's assumed role in the MAP but it also goes beyond the CAP's assumption that the role of the customer is that of an inventive user who passes on his ideas, and perhaps designs and implementation information freely or at minimum cost to the manufacturer who benefits from its economic dissemination. The entrepreneurial user, the industrial consumer who becomes a supplier, is not properly accounted for in the literature of organizational buying behavior.

Particularly interesting is the active mode of user-initiated product innovation in which the user who has applied his self-invented device as a process innovation gains from its general dissemination as a product innovation (UII3). Elements of all three approaches discussed above are required to account for this. The appropriability of benefit hypothesis suggests that UII3 will be most improbable: the innovator cannot benefit for long from non-embodied knowledge and has no other means of establishing the quasi-monopoly which confers response time. The hypothesis does not exclude entirely the possibility that a firm-level quasi-monopoly may prevail temporarily as a result of trade know-how, but does now explain why UII3 should endure as a form of manufacturer-user interaction from which the parties benefit to their mutual satisfaction. It does not, therefore, assist in the prediction of where and when UII3 will occur. For this task, the appropriability regimes and transaction costs hypotheses are necessary. The former stresses the importance of complementary assets while the latter indicates that manufacturer-opportunism can be forestalled by the development of transaction-specific assets upon which the manufacturer depends for response time.

The transactions costs approach generates general explanations of the observed phenomena while the alternative hypotheses examined refer to specific instances of user- or manufacturer-innovation which are governed by particular transactions costs considerations (identifying specific sources of opportunism, for instance, in the patenting problems of the would-be innovator). Both the general and particular explanations are valuable. The transactions cost approach is not easily operationalized and other hypotheses are useful to illustrate specific sources of opportunism and asset specificity, and the circumstances in which they influence the functional locus of innovation.

REFERENCES

1 E A von Hippel, Successful industrial products from customer ideas, J Market, 42(1) (1978) 39-49.

2 G R Foxall, A conceptual extension of the customer-active paradigm, Technovation 2 (1986) 17-27.

3 G R Foxall and B Johnston, Strategies of user-initiated product innovation, Technovation (in press).

4 K J Blois, Vertical quasi-integration, J. Indus. Econ, 20 (1972) 253-272.

5 G R Foxall and J D Tierney, From CAP1 to CAP2: User-initiated innovation from the user's point of view, Man. Dec, 22(5) (1984) 3-15.

6 G R Foxall, F S Murphy and J D Tierney, Market development in practice: A case study of user-initiated innovation, J. Market. Manage, 1(3) (1985), 201-211.

7 G R Foxall, B Johnston and F S Murphy, The development of control software by British Aerospace, Warton: A case history of user-initiated product innovation, J. Market. Manage, 2(3) (1987) 259-74.

8 E A von Hippel, Appropriability of innovation benefit as a predictor of the source of innovation, Res. Pol. 11 (1982) 95-115.

9 D J Teece, Profiting from technological innovation: Implications for integration, collaboration, licensing and public policy, Res. Pol, 15 (1986) 285-305.

10 W J Abernathy, and J M Utterback, Patterns of industrial innovation, Technol. Rev, 80(1) (1978) 40-47.

11 O E Williamson, The vertical integration of production: Market failure considerations, Amer. Econ. Rev, 61 (1971) 112-123.

12 O E Williamson, Markets and Hierarchies: Analysis and Antitrust Implications, Free Press, New York, 1975.

13 O E Williamson, The modern corporation: Origins, evolution, attributes, J. Econ. Lit, 19 (1982) 1537-1568.

14 S Globerman, Markets, hierarchies and innovation, J. Econ. Issues, 14 (1980) 265-288.

15 K Monteverde and D J Teece, Appropriable rents and quasi-vertical integration, J Law and Econ, 25 (1982) 321-328.

16 E A von Hippel, Lead users: A source of novel product concepts, Man. Sci, 32 (1986) 791-805.

Innovation: Adaptation and Growth, edited by R. Rothwell and J. Bessant
Elsevier Science Publishers B.V., Amsterdam, 1987 — Printed in The Netherlands

INTELLIGENT CORPORATE STRATEGY FOR NEW PRODUCT DEVELOPMENT CONSIDERING GROWTHABILITY AND ADAPTABILITY

TANAKA YOSHIHIKO

Waseda University, Graduate School and School of Sci & Engg,

Dept of IE & ME, Tokyo, Japan

In this paper, the intelligent corporate strategy for new product development is expained. Namely, the intelligence amplification criteria is made as a draft. And then, the special factor analysis method and the special torm system equation method as its methodologies or tools are created.

As a result, the effectiveness of the proposed methods and tools is cleared up and the knowledge for the basis of the industry from now to the 21st century is constructed.

1. Introduction

The authors have been going to research and give some lectures regarding the diagonosis, analysis and design methods of the intelligent administration strategy system(IASs), the intelligent business management system(IBMs) and the intelligent CAD/CAM/CAT operating system(ICOs) under the practical age of the utilization for the universal space and the diffusion of construction of the global intelligent freedomism(GIF) system as the construction of new ABC power through proper and fair competition and cooperation of private and public enterprises.

On these IASs, IBMs and ICOs, for growth and adaptation with capital, customers, emploee, correlated enterprises and public, it is necessary to develop timly new products and to determine of merger and acquisition or partition of the corporations through reviewing and recreating the criteria which satisfys with administration goal and environment togather with utilization of various kinds of resouces effectively for capital amplification and the elevation of living standerds for individual.

For example, Bayer AG which is one of the largest chemical and health care companies in West Germany introduced a new organizational structure in 1984 which responses to the changes in both the internal and external influences on the corporation for new growth and adaptation.

The Corporate Policy Group in Oxford, UK, and in MIT, USA point out the importance of the intelligent corporate strategy which is grasped as the concept of "intelligent corporate governance" and the importance of the beneficial aspect of new products or new investment and new technology through activation of inherent technology on each enterprise in parallel with costing aspect of them in large scale corporations. Because, the ordinary long-range planning and strategy formulation will, oftenly, be nothing more than a paper exersise, therefore, in more limited sense, it is necessary to construct the valuable strategy both in name and reality.

And in Japan, as the country of the first creditor, Japanese government under the guidance of new wisdomization organon has determined the reformation of administration style for JNR(Japanese National Railways) and began to plan the introduction of new taxation system for growth, adaptation, enrichment and intensification of Japanese economy and world-wide economy, both norminally and virtually, through new science and technology policy, trade and commerce policy, defence policy, diplomacy policy, education policy and finance policy. Therefore, many superior Japanese corporations are searching for new intelligent corporate strategy(NICS), constructing of the needs assessment technology organization(NATO) and the determination of management evaluation system(DOMES) on the basis of the objective and scientific observation and consideration in order to correspond with continuation of corporations.

In this paper, the authors explain the criteria for the determination of strategy regarding new product development and merger and aquisition or partition of corporations under this background and its methodology. That is, the intelligence amplification criteria, namly, wisdomization criteria as the criteria for such determination is made as a draft, and, the special factor analysis method and the special torm system equation method using the supercubic for the administraction of the new products clustered as the "stars " as its methodologies are created.

2. The needs of GIF and its methodology as the corporate strategy for new product innovation

According to the maturization and the advancement of the industry, the administrative objects are naturally into global because of the expansion of the administration scale, the superiority of the product ability and quality and the accumulation of capital. Because, the superior and required products are naturally disseminated or affect broad domain deeply, therefore, their administrative activities get into global,complex and multiple networked circulation activities. Fig. 1 shows the scheme diagram of such relationship of activities.

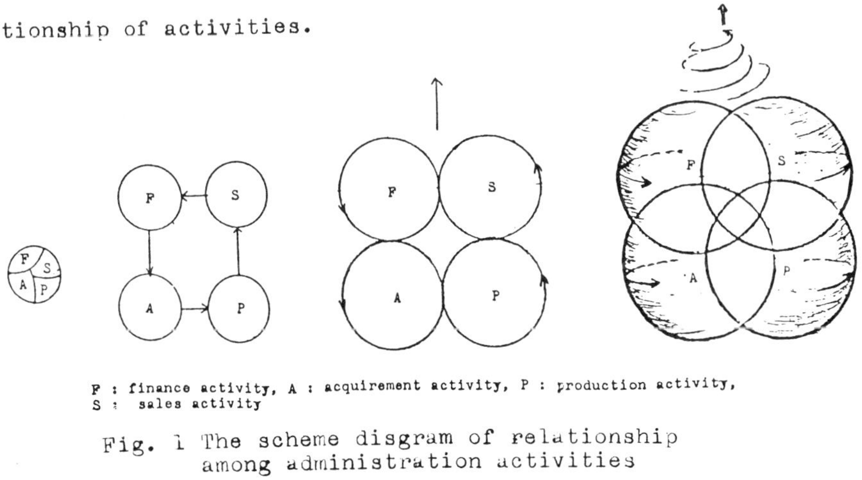

F : finance activity, A : acquirement activity, P : production activity,
S : sales activity

Fig. 1 The scheme disgram of relationship among administration activities

For example, some representative corporations in Japan, today, are intensifying such role and many countries in the world require strongly that many Japanese corporations play such a role. And then, some Japanese top managements are proposing the importance of the change from the introduction of the local rule to the introduction of the international rule which is based on the logic of capital and the review of the Anti-Monopoly Law, the Commercial Law and the Corporation Tax Law including the stock option. Then, as the PIMS project group at Columbia University has pointed out, for the continuous property of the corporation, for the fulfilment of the strong leadership on market and for the construction of intelligent cash flow, it is necessary to furnish the speciality, the essential function or the essential utility as superior products and it means the importance of assessment not only for the cost aspect but also for gain aspect of products.

These facts mean the importance of the intelligence amplification including a profit margine amplification as corporation through the global intelligent freedomism(GIF) system for flexibility and elevation of individual living standard. For that purpose, the methodology which analyzes the speciality and activates into design is needed. The special supercubic factor analysis method and the special torm system equation method using the supercubic explained in this paper are the methodologies in accordance with such needs.

3. Modeling and methodology

3.1 The generalized information space and the generalized factor analysis method

Generally, on the global corporate administration, the administration information for corporate strategy is versatile and mass volumed. Such administration and management information is defined as the generalized superinformation space and expressed as the eq

(1).

$$\overset{\circ}{X}^{\omega} = [\ X^{\omega}\ ]^{\omega} \qquad (1)$$

where

$$X^{\omega} = \langle\ x_{i_1 i_2 \ldots i_\omega}(l)\ \rangle$$

$$x_{i_1 i_2 \ldots i_\omega}(l) = \begin{pmatrix} x_{i_1}(l) \\ \vdots \\ x_{i_j}(l) \\ \vdots \\ x_{i_\omega}(l) \end{pmatrix}$$

$$x_{i_j}(l) = (\ x_{1_j}(l), x_{2_j}(l), \ldots, x_{p_j}(l)\ )$$

$$\begin{cases} \sum_{i_1 i_2 \cdots i_\omega} x_{i_1}(l) x_{i_2}(l) \ldots x_{i_\omega}(l) = 0 & \text{(if each variable vector axis } i_j \text{ is indeoendent)} \\ \sum_{i_1 i_2 \cdots i_\omega} x_{i_1}(l) x_{i_2}(l) \ldots x_{i_\omega}(l) = \eta_{i_1 i_2 \ldots i_\omega}(l) & \text{(if the co-relationality exists)} \end{cases}$$

l ; the individual or the observation object

i_j ; the variable vector axis

$\overset{\circ}{X}^{\omega}$; the generalized superinformation space

For the generalized superinformation space discribed above, the generalized factor analysis method conducts the following discomposition, and then, wisdomizes the residual super ω-space E^{ω} under some criteria while minimizing the number of common factors.

$$\overset{\circ}{X}^{\omega} = \overset{\circ}{F}^{\omega} \cdot \overset{\circ}{A}^{\omega\prime} + \overset{\circ}{E}^{\omega} \qquad (2)$$

$\overset{\circ}{F}^{\omega}$; the factor score super ω-space (the common factor)

$\overset{\circ}{A}^{\omega}$; the factor load super ω-space (the special factor)

$\overset{\circ}{E}^{\omega}$; the residual super ω-space

So, the omega product spac $\overset{\circ}{\Omega}^{\omega}$ is expressed as follows.

$$\begin{aligned} \overset{\circ}{\Omega}^{\omega} &= (\overset{\circ}{E}^{\omega})^{\omega} \\ &= \sum_{r=0}^{\omega} \binom{\omega}{r} \overset{\circ}{X}^{\omega\,\omega-r} (-\overset{\circ}{F}^{\omega} \cdot \overset{\circ}{A}^{\omega\prime})^{r} \qquad (3) \end{aligned}$$

If the following criteria is created as the determination criteria of $\overset{\circ}{F}^{\omega}$ and $\overset{\circ}{A}^{\omega}$,

$$\phi = \mathring{1}^{\omega} - \frac{tr^{\omega}\mathring{\psi}^{\omega}}{tr^{\omega}\mathring{G}^{\omega}} \longrightarrow \text{maximize} \qquad (4)$$

where

$$\mathring{\psi}^{\omega} = (\mathring{G}-\mathring{Q}(\mathring{A}^{\omega})^{\omega}\mathring{Q})^{\omega}$$

$$\mathring{G}^{\omega} = (\mathring{Q}(\mathring{O}^{\omega}-\theta\mathring{D}_W)\mathring{Q})^{\omega}$$

$$\mathring{\psi} = \mathring{Q}(\mathring{W}-\theta\mathring{D}_W)\mathring{Q}$$

$$\mathring{W} = \mathring{O}^{\omega}-(\mathring{A}^{\omega})^{\omega}$$

$$\mathring{O} = \mathring{X}^{\omega}$$

$$\mathring{D}_W = \text{diag}\ \mathring{W}$$

then, the optimal $\mathring{\hat{A}}^{\omega}$ may satisfy following equation.

$$(\mathring{G}-\mathring{Q}(\mathring{A}^{\omega})^{\omega}\mathring{Q})^{\omega -1}\mathring{Q}(\mathring{A}^{\omega})^{\omega -1}\mathring{Q} = \mathring{O} \qquad (5)$$

Using the $\mathring{\hat{A}}^{\omega}$, the $\mathring{\hat{F}}^{\omega}$ is determined.

3.2 The special supercubic factor analysis method

On the new product strategy under the administration environment described above, the special supercubic factor analysis method which is tripled of the special supercubic principle component method and special supercubic clustering method and is for the more definitive use in the more limited sense is effective and efficient.

That is, on the special supercubic factor analysis method, the administrative information supercubic clustered through the special supercubic principle component and the special supercubic clustering method on the basis of the latent rating structure model, such as, by market, by product and by cross of market and product, is discomposed as eq (6). On that occasion, the number of common factors is minimized, the special factor is wisdomized under some criteria and the new product strategy is determined through the torm system equation analysis method for the factor load supercubic.

$$\mathring{\mathbf{X}} = \mathring{\mathbf{F}}\cdot\mathring{\mathbf{A}}' + \mathring{\mathbf{E}} \qquad (6)$$

where

$$\mathring{\mathcal{X}} = [\{X_{ijk}\}] = \begin{bmatrix} X_{n1n} & \cdots & X_{nnn} \\ \vdots & & \vdots \\ X_{n11} & \cdots & X_{nn1} \end{bmatrix} \quad \begin{bmatrix} X_{11n} & \cdots & X_{1nn} \\ \vdots & & \vdots \\ X_{111} & & X_{1n1} \end{bmatrix}$$

$\mathring{\mathcal{X}}$; the administrative information supercubic

$\mathcal{X}_{ijk}$; the subcubic $\mathcal{X}$ at portion ijk

If the following criteria on the basis of the square of the residual supercubic using the correlation supercubic $\mathring{R}$ and the trirelationality supercubic $\mathring{T}$,

$$\phi = \mathring{1} - \frac{\mathrm{tr}\,\mathring{\Psi}^2}{\mathrm{tr}\,\mathring{G}^2} \longrightarrow \text{maximize} \qquad (7)$$

then,

$\tilde{\mathring{A}} = \mathring{Q}\cdot\mathring{A}$ (when $\mathring{R}$ is used)

$\mathring{A} = \mathring{A}^{-1}\cdot\mathring{Q}^{-1}\cdot\mathring{G}\cdot\mathring{Q}'^{-1}\cdot\mathring{A}'^{-1}$ (when $\mathring{T}$ is used)

Fig. 2 shows the procedure of the special supercubic factor analysis method.

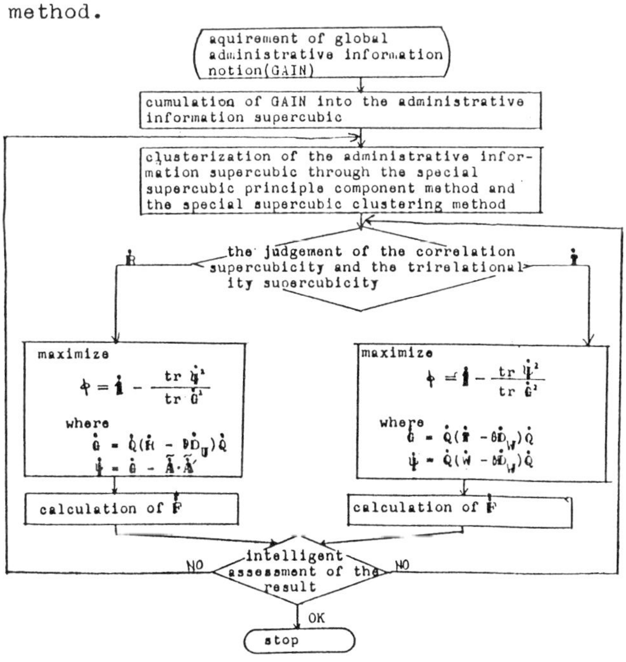

Fig. 2 The procedure of the special supercubic factor analysis method

3.3 The supercubic torm system equation method

When the factor load supercubic and the factor score supercubic are stochastic, the dynamic assessment for the factor load supercubic have been got through the special supercubic factor analysis method is performed by the formulation (8).

$$\|\mathring{A}_{\mathring{n}\in\mathring{N}}(t)\|^{\sim} = \|\mathring{\Omega}(t-1)\|^{\sim}\|\mathring{A}_{\mathring{n}\in\mathring{N}}(t-1)\|^{\sim} + \|\mathring{\phi}(t-1)\|^{\sim}\|\mathring{U}(t-1)\|^{\sim} + \|\mathring{E}(t-1)\|^{\sim} \qquad (8)$$

where

$\mathring{A}_{\mathring{n}\in\mathring{N}}(t)$; the stochastic factor load supercubic

$\mathring{U}(t-1)$; the wisdomization supercubic

$\mathring{E}(t-1)$; the residual supercubic at period t-1

$\mathring{\Omega}(t-1), \mathring{\phi}(t-1)$; the transformation supercubic

$\|\cdot\|^{\sim}$; the torm of the stochastic supercubic .

Solving the eq (8) through the Laplace Transformation method, the following result is got.

$$\|\mathring{A}_{\mathring{n}\in\mathring{N}}(t)\|^{\sim} = \mathcal{L}^{-1}\{\|\mathring{A}_{\mathring{n}\in\mathring{N}}[s]\|^{\sim}\}$$

where

$$\|\mathring{A}_{\mathring{n}\in\mathring{N}}[s]\|^{\sim} = (1 - e^{-s\frac{D_1}{s}} - e^{-s\frac{D_2}{s}})^{-1}\Big[e^{-s}\Big\{\|\mathring{\Omega}(0)\|^{\sim}\Big(\frac{1}{s}\|\mathring{A}(0)\|^{\sim} + \frac{C_1}{s^2}\Big) + \frac{D_1 C_1}{s^3}\Big\} + e^{-s}\Big\{\|\mathring{\phi}(0)\|^{\sim}\Big(\frac{1}{s}\|\mathring{U}(0)\|^{\sim} + \frac{C_2}{s^2}\Big) + \frac{D_2 C_2}{s^3}\Big\} + e^{-s\frac{D_2}{s}}\|\mathring{U}[s]\|^{\sim}\Big]$$

Fig.3 shows the draft for the design of the supercubic torm system equation analyzer on the basis of the analysis result and the analysis method described above.

Using the supercubic torm system equation analyzer which consists of the small intelligent NMR unit, ie, "The Thinkman", the fs order absorption spectorscope unit, the optocommunication unit, the control unit of the large rocket engine and the laser gyro uint as new product in the large scale corporation strategy, the new product strategy towards the 21st century shall be prospective.

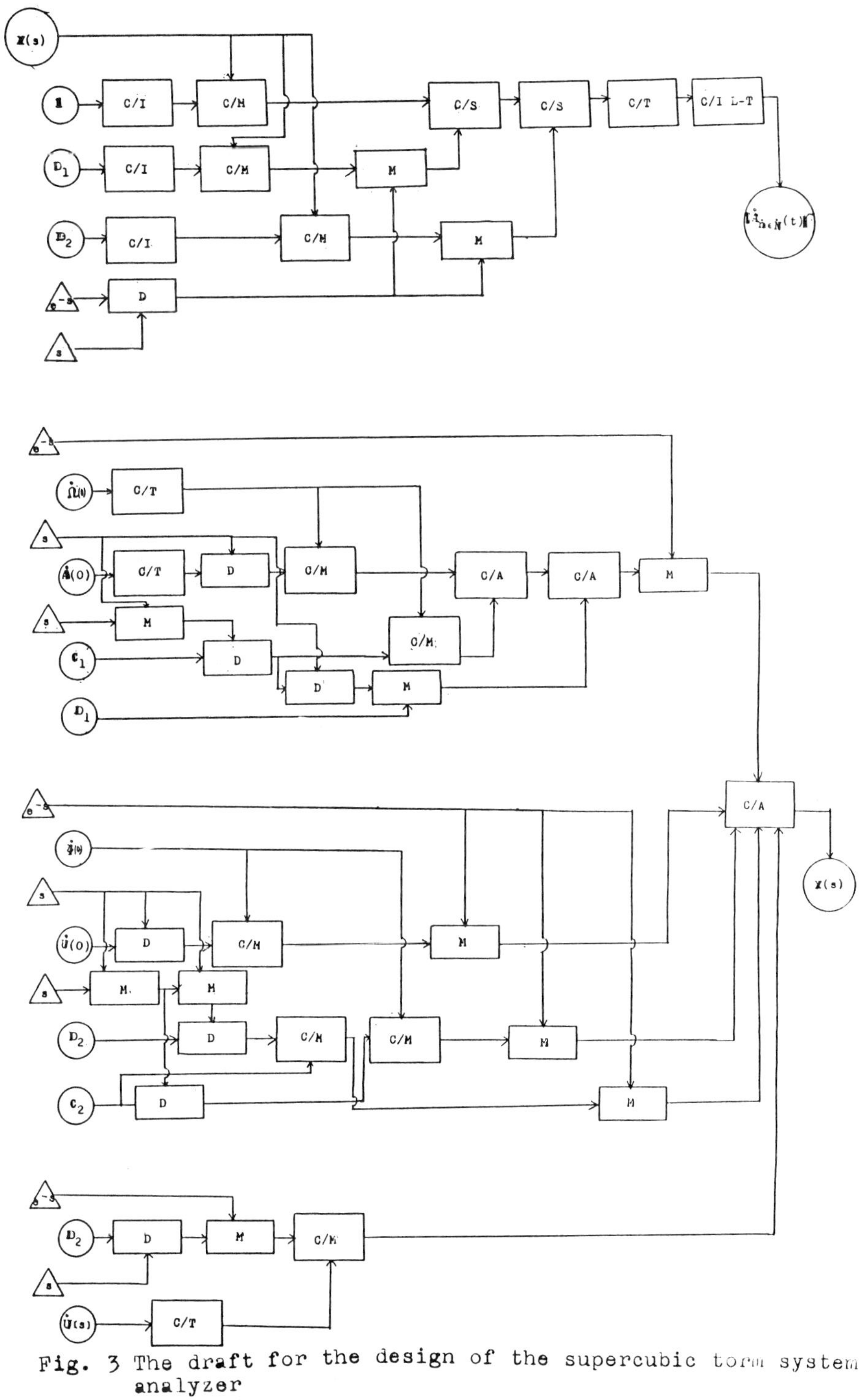

Fig. 3 The draft for the design of the supercubic form system analyzer

4. Some cases of the new product development through the supercubic administration strategy method

Tab 1 shows the some representative new products developed in 1986 through the supercubic administration strategy method and the intelligent supercubic assessment method(ISAM).

And then, those products, developing process, organizing system and policy·promotion and evaluation mechanizm·consist of the basis of the international basic reseach mechanizm, such as, the human frontier science program which was discussed through Venetia Summit on June, 1987.

Tab 1 Examples of new products developed in 1986 through the supercubic administration strategy method and the intelligent gent supercubic assessment method

New Products	Development Organization	Features
Engine for HI Rocket (LE 5)	·Mitsubishi Heavy Ind. Corp. ·IHI Corp. ·Space Development Corp.	·The second stage rocket engine, HI Rocket, which uses the liquid oxygen and the liquid hydrogen as the propulsion fuel ·The ability for the shoot-up of the static satellite is 550kg. ·The product isaintegrated outcome which consists of the several high level rocket technology, for example, the turbo-pump, thrust room and the gass generator.
small VTR camera(GR-C7)	Nihon Victor Corp.	·The weight of the main body is 1.3kg. ·Using of the compact cassette tape for smallization ·Reduction of the number of parts by the introduction of the IC technology
composite urban development (Ark hills)	Mori building Corp.	·The composite urban which consists of the super high office building, housing, city-typed hotel, television stadio and concert hall ·The product is able to utilize the television conference system and a high speed degital circuit network.
home copy (Family Copia FC 3)	Canon Corp.	·Integration of the exposure drum and tonor into super small cartridge ·The weight is 13kg. ·The price is 99,800 yen.
interest linkage bond	Nomura Securities Corp.	·The low risk high return typed bond for investors aiming at capital gain ·The product is based on a yield basis of 30 years national bond in USA as base market return.

5. Conclusion

In this study, we have explained the strategy of new product development. Namely, we have created the intelligence amplification criteria, ie, wisdomization criteria and the special factor analysis method and the special torm system equation method as its methodologies.

As a result, we have got the following knowledge and given the following contribution.

i) On the corporate strategy for new product development, the proposed intelligence amplification criteria, ie, wisdomization criteria is effective and efficient. Especially, for the new round regarding the global intelligent freedomism(GIF) system towards the 21st century, it plays a important role.

ii) On that occasion, the proposed methods, ie, the special factor analysis method and the special torm system equation method are able to apply not only to the develoopment of medical treatment units, new medicine, semiconductors, life science materials, sports materials, optoelectronics science materials, new energy, ocean developing, aerospace and financial and insurance products, namely, the products belonging to the clusters "the stars" or "the wildcats" which are categorized through the expanded analysis by the strategy research group at Columbia University Graduate School of Business, but also to the development of automobile, iron and steel, housing, living materials, traffic and closing, namely, the products belonging to the clusters "the caws" or "the dogs".

iii) By linkage of the proposed methods or plant to the trispectrum supercubic and the coherency supercubic units which use the fs(10^{-15}sec) order laser beam, the effectiveness of the application for the highest leading technology industry increases nonlinearly.

iv) From those the results, through repleting the softwares and hardwares regarding the proposed principles and methods as the kernel technologies, the corporate strategy for new product development shall be worth-while from now to the 21st century.

REFERENCES

1 Muramatsu,R.(1979), 'The basic principle of production management' Kunimoto Shobo, (Japan)

2 Tanaka,Y.(1986),"Designing and analysis of production management system for construction of large scale system through strategic decision supported intelligence", Proceeding of the 4th IFAC/IFORS Symposium, Zürich,(Switzerland),667-672

3 Muramatsu,R.,(1985),"Some ways to increasing flexiblity in production systems", Int.J. of Prod.Res. (UK),Vol.23,No.4,691-704

4 Tanaka,Y.,(1986),"A Proposal for a new product innovating method through fusion strategy in intelligent CAM(ICAM)", J. of Technovator(Netherland)

5 Tanaka,Y.,(1985),"A proposal and analysis of tri-relational forecasting and control method",Proc. of the 8th ICPR(WG),127-32

6 Tanaka,Y.,(1985),"An analysis of the multi-product cross-correlation forecasting method for designing a flexible computer integrated manufacturing(CIM) system",Proc of Conputer & IE,(USA),27-31

7 Tanaka,Y.,(1985),"A proposal for flexible automatic testing system through multi-dimensional nonparametric management system", AMSE Review,(France),Vol.1,No.4,25-30

8 Tanaka,Y.,(1984),"A proposal for flexible automatic process design at multi-products assembly line system",Proc. of Int. AMSE Conf.,(USA),Vol.4,121-131

9 Tanaka,Y.,etal.,(1986),"Analysis and designing of intelligent business and management system based on microelectronics", Proc. of IMC,(Japan),314-319

10 Tanaka,Y.,etal.,(1986),"Design for the intelligent computer integrated manufacturing system through the cubic Fourier transformation method based on the Tit element",Proc. of the 36-th Electronics Components Conference,(USA)

11 Salvendy,G.,(1986),'Handbook of industrial engineering',John Wily & Sons,(USA)

12 Mital,A.,(1985),"Use of anthropometry and dynamic strength in developing screening and placement procedures for workers" Proc. of the 8th ICPR(WG),903-909

13 Bullinger,H.J.,etal.,(1985),"Toward the factory of the future", Proc. of the 8th ICPR,(WG)

14 Tanaka,Y.,(1986),"Development and design method of the intelligent robot", Materials of 9th Annal British Robot Association Conf.(BRA-9),(UK)

15 Muramatsu,R.,(1980),'Production Management System', Asakura-Shoten,(Japan)

16 Hambrick,D.C.,et al, "Strategic attributes and performance in the BCG Matrix——A PIMS-based analysis industrial product businesses",Academy of management Journal,(1982) Vol.25,No.3,510

17 Vossberg,H.,(1985),"Bayer reorganizes in response to growth",Long Range Planning,Vol.18,No.16,18

Innovation: Adaptation and Growth, edited by R. Rothwell and J. Bessant
Elsevier Science Publishers B.V., Amsterdam, 1987 — Printed in The Netherlands

DanProd

A model for assessment of new product ideas and projects.

Jens H. Arleth, M.Sc., B.Com.
Senior Consultant, Dpt. of Industrial Management
Technological Institute, 2630 Taastrup, Denmark
This paper is published by permission of the Danish Technological Institute.

Abstract

The DanProd model is a decision tool for evaluation of new product ideas and concepts. DanProd is normally used in the early stages of the development process.

The DanProd model has been developed by the Dept. of Industrial Management, Technological Institute, in cooperation with Robert G. Cooper of MacMaster University, Toronto. The basis of the development work is one of detailed analyses of 195 Canadian and 26 Danish product development cases out of which one half came out successfully, and the other half as failures. In each and every case, it is a question of products fully developed and on the market for at least two years.

This paper describes in brief what DanProd is. Furthermore the use of DanProd as a management tool is illustrated by an example.

What is DanProd

The DanProd model will predict and state the reasons for the chances of success for a new product development project. This model is available as software for the IBM Personal Computer.
When applying DanProd, evaluators selected within the company will fill in a questionnaire of 48 questions which research has proven to be essential. On the basis of the answers. a profile is drawn up of the projects to be evaluated. This profile will be compared to the profiles of the 221 projects integrated in DanProd. On the strength of this, the succes probability is calculated by percentage, and the

factors contributing positively or negatively towards the chances of success of the project being considered by the particular company will be stated.

The forerunner of DanProd is the Canadian NewProd, developed by Cooper, used already for several years by American and Canadian corporations. As from the middle of 1984, NewProd has been tested successfully with a considerable number of companies in Denmark. Thus, we have gained varied experience allowing for a further development of NewProd. The result of this is a particular Danish model, named DanProd.

An outstanding quality of this method proved to be that it does, in a highly rewarding manner, direct the discussion of the dicisionmaker towards the most essential and critical points of the particilar project, incl.:

- What is the scope of the new project
- What are the drawbacks of the new project
- On which points is there a considerable amount of incertainty about the projects

This analysis is used to set up a diagnosis and to devise a plan of action for the further process, incl.:

- Does this project offer such a high degree og success probability that it would seem reasonable to pursue it further?
- Should it be checked whether the positive aspects of the project will actually come true?
- What are the negative aspects for this project, and could we improve on these aspects and thus enhance the chances of success?
- To which extent may further information have to be compiled to enable us to decide upon the final destiny of the project?

The following pages show copies of a DanProd analysis printout from the company S&W Medico Teknik A/S. A review is made in the following sections of what may be read from this printout, and of how DanProd may be applied as a link in the company decisions on new product development projects.

DanProd Printout, Page 1 .

DANPROD - IDEA ASSESSMENT

Project: Rescar3 Date: 05-22-1987

DPT. INDUSTRIAL MANAGEMENT - TI

PROBABILITY OF SUCCESS IS 75%

The project Rescar3
is expected to be a success
with a likelihood of 75%

PROBABILITIES

1	HRA	R&D Manager	74%
2	R.J	Service mng.	68%
3	IOG	Project manager	79%
4	DOR	Sales Director	86%
5	O.S	Produktion Director	84%
6	NID	Marketing Director	44%
7	J.L	Product manager	91%
8	U.L	Q.A. manager	50%
WEIGHTED MEAN			75%

PROJECT PROFILE

1 Product Superiority/Quality

2 Economic Advtg to user

3 Market Need, Growth & Size

4 Market Competitiveness

5 First into market

6 Overall Company-project fit

7 Technological synergy

8 Market newness to company

9 Project Determinateness

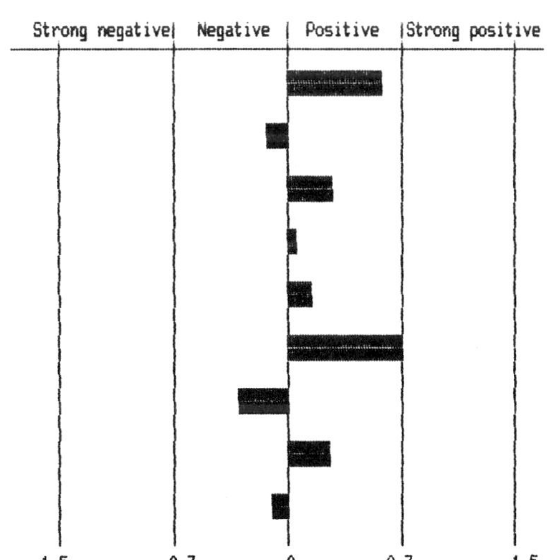

DanProd Printout Page 2

EVALUTORS INPUT

	EVALUATOR	HRA		R.J		IOG		DOR		O.S		NID		J.L		U.L		WT'D MEAN, 8 EVALUAT		
		R	C	R	C	R	C	R	C	R	C	R	C	R	C	R	C	RAT.	STD.DEV	CON
1	Finances o.k.	10	10	10	8	10	10	9	9	9	9	10	10	8	8	10	10	9.5	0.7	9.3
2	R&D skills o.k.	8	10	6	6	6	10	6	10	5	9	2	10	7	8	7	9	5.8	1.8	9.0
*** 3**	**Engineering o.k.**	**8**	**10**	**3**	**8**	**5**	**10**	**1**	**9**	**3**	**9**	**2**	**10**	**9**	**7**	**9**	**9**	**4.9**	**3.0**	**9.0**
4	Mktg research o.k.	5	5	9	8	10	10	9	9	6	9	9	8	10	10	7	8	8.4	1.7	8.4
5	Management o.k.	10	10	8	10	10	10	10	10	9	9	9	9	10	9	9	9	9.4	0.7	9.5
*** 6**	**Production o.k.**	**10**	**10**	**0**	**9**	**10**	**10**	**2**	**9**	**4**	**7**	**8**	**8**	**8**	**7**	**5**	**10**	**5.9**	**3.6**	**8.8**
7	Saleforce/distr. o.k	10	10	8	10	10	10	10	10	10	10	7	10	10	10	8	8	9.2	1.2	9.8
8	Advert/promo o.k.	10	10	8	9	10	10	8	10	10	10	8	8	10	10	8	9	9.1	1.0	9.5
*** 9**	**Innovativeness**	**4**	**8**	**1**	**9**	**0**	**10**	**6**	**10**	**4**	**7**	**0**	**8**	**2**	**8**	**8**	**8**	**3.1**	**2.7**	**8.5**
10	Hi tech product	0	8	2	9	4	8	5	10	2	9	0	10	1	8	2	9	2.0	1.7	8.9
11	Big ticket product	5	10	9	9	5	8	6	10	5	0	7	5	5	9	3	7	5.7	1.7	7.3
12	Technically complex	2	7	5	8	6	8	6	10	9	9	5	8	5	8	3	8	5.3	2.0	8.3
13	Market idea	9	6	8	8	5	6	9	10	9	9	10	10	8	10	7	7	8.3	1.3	8.3
***14**	**Clear product specs**	**9**	**8**	**7**	**8**	**2**	**10**	**7**	**10**	**3**	**9**	**8**	**9**	**5**	**7**	**7**	**7**	**5.9**	**2.4**	**8.5**
15	Clear tech solution	2	7	8	8	3	8	5	10	2	9	3	8	3	9	5	8	3.9	1.9	8.4
***16**	**Custom product**	**8**	**4**	**2**	**7**	**5**	**5**	**9**	**10**	**5**	**9**	**5**	**10**	**4**	**9**	**3**	**8**	**5.1**	**2.2**	**7.8**
17	Defensive product	4	5	5	8	3	10	2	10	4	8	4	8	2	9	5	8	3.5	1.1	8.3
18	Relative investment	0	10	6	8	5	8	5	10	4	8	4	9	6	8	3	8	4.0	1.9	8.6
19	New customers	0	10	4	8	0	10	1	10	0	10	0	10	1	10	2	4	0.8	1.3	9.0
20	New product class	0	10	0	10	0	10	0	10	0	10	0	10	0	10	0	10	0.0	0.0	10.0
21	New needs	0	10	0	10	0	1	0	10	0	10	0	10	3	9	0	10	0.4	1.0	8.8
***22**	**New production**	**0**	**10**	**8**	**8**	**0**	**10**	**8**	**10**	**0**	**10**	**3**	**8**	**2**	**9**	**1**	**10**	**2.6**	**3.2**	**9.4**
23	New R and D	0	10	2	9	0	10	6	10	0	10	1	8	2	9	0	10	1.4	2.0	9.5
24	New salesf/distr.	0	10	1	9	0	10	3	10	0	10	2	6	0	10	0	10	0.7	1.1	9.4
25	New advert/promo	0	10	1	9	0	10	5	10	0	10	4	7	0	10	0	10	1.1	1.9	9.5
26	New competitors	0	10	1	9	0	10	2	10	0	10	0	10	0	10	0	9	0.4	0.7	9.8
27	Unique benefits/feat	8	7	8	8	7	7	9	10	8	10	7	9	9	9	8	8	8.0	0.7	8.5
28	Superior product	8	7	9	8	8	8	9	10	10	9	7	7	9	10	8	7	8.6	0.9	8.3
***29**	**Reduce cust costs**	**5**	**5**	**7**	**7**	**5**	**6**	**6**	**10**	**8**	**9**	**0**	**10**	**5**	**5**	**1**	**8**	**4.5**	**2.8**	**7.5**
***30**	**Unique task**	**5**	**5**	**2**	**8**	**8**	**8**	**9**	**10**	**1**	**9**	**6**	**6**	**3**	**9**	**7**	**7**	**5.1**	**2.9**	**7.8**
31	Product quality	5	5	7	8	8	9	5	10	8	9	9	8	5	5	5	7	6.7	1.6	7.6
32	Relative price	5	5	5	7	6	6	5	10	3	8	5	9	3	5	5	7	4.6	0.9	7.1
***33**	**First in**	**5**	**5**	**1**	**8**	**0**	**10**	**8**	**10**	**1**	**9**	**2**	**10**	**5**	**7**	**3**	**8**	**3.0**	**2.6**	**8.4**
34	Mass market	5	5	7	7	7	9	9	10	4	8	7	8	8	8	5	8	6.7	1.6	7.9
***35**	**Customer need**	**5**	**8**	**9**	**7**	**10**	**10**	**10**	**10**	**10**	**10**	**9**	**10**	**10**	**10**	**4**	**7**	**8.6**	**2.2**	**9.0**
36	Potential demand	0	10	2	7	0	10	0	10	0	9	0	10	0	10	3	7	0.5	1.0	9.1
***37**	**Market size**	**5**	**5**	**3**	**9**	**7**	**8**	**10**	**10**	**7**	**9**	**6**	**8**	**8**	**9**	**5**	**8**	**6.5**	**2.1**	**8.3**
***38**	**Market growth**	**5**	**5**	**3**	**8**	**5**	**8**	**9**	**10**	**5**	**8**	**5**	**7**	**10**	**10**	**2**	**8**	**5.8**	**2.7**	**8.0**
***39**	**Product homogeneity**	**4**	**5**	**2**	**9**	**0**	**10**	**5**	**10**	**7**	**8**	**4**	**6**	**5**	**8**	**1**	**7**	**3.4**	**2.3**	**7.9**
***40**	**Competitive market**	**5**	**2**	**5**	**6**	**10**	**10**	**7**	**10**	**9**	**9**	**7**	**9**	**2**	**9**	**2**	**7**	**6.2**	**2.9**	**7.8**
***41**	**Price competition**	**5**	**2**	**5**	**6**	**7**	**8**	**6**	**10**	**8**	**9**	**4**	**4**	**0**	**10**	**5**	**7**	**5.0**	**2.6**	**7.0**
***42**	**Many competitors**	**5**	**2**	**5**	**6**	**6**	**8**	**5**	**10**	**8**	**10**	**10**	**10**	**2**	**9**	**2**	**9**	**5.5**	**2.8**	**8.0**
***43**	**Dominant competitor**	**5**	**2**	**6**	**7**	**6**	**7**	**4**	**10**	**8**	**10**	**6**	**10**	**2**	**9**	**2**	**8**	**4.9**	**2.1**	**7.9**
44	Competitor loyalty	5	2	6	7	3	8	1	10	5	5	0	10	2	9	1	8	2.3	2.0	7.4
45	Cust satisfaction	4	2	4	6	3	8	3	10	3	9	8	9	4	5	5	5	4.3	1.8	6.8
***46**	**Many intros**	**3**	**6**	**5**	**7**	**8**	**8**	**6**	**10**	**3**	**9**	**5**	**9**	**0**	**9**	**3**	**6**	**4.2**	**2.3**	**8.0**
***47**	**Dynamic needs**	**0**	**9**	**3**	**7**	**0**	**10**	**8**	**10**	**6**	**9**	**1**	**9**	**1**	**8**	**1**	**8**	**2.6**	**2.9**	**8.8**
***48**	**Government role**	**1**	**8**	**2**	**8**	**10**	**10**	**10**	**10**	**9**	**10**	**4**	**6**	**10**	**9**	**8**	**8**	**7.1**	**3.5**	**8.6**

NOTE: The R column gives the ratings and the C column the confidences for evaluators.
High disagreement questions are emphasized and marked by an *

DanProd Printout, Page 3

QUESTIONS WITH HIGH DISAGREEMENTS
The following questions had high disagreements among evaluators. Evaluators should compare and discuss their ratings.

		WDT MEAN	STD DEV
6	Production o.k.	5.94	3.56
48	Government role	7.13	3.51
22	New production	2.61	3.18
3	Engineering o.k.	4.92	3.00
47	Dynamic needs	2.57	2.90
40	Competitive market	6.23	2.90
30	Unique task	5.10	2.89
29	Reduce cust costs	4.48	2.83
42	Many competitors	5.53	2.77
9	Innovativeness	3.07	2.74
38	Market growth	5.78	2.72
33	First in	3.00	2.64
41	Price competition	4.98	2.59
14	Clear product specs	5.90	2.36
46	Many intros	4.17	2.33
39	Product homogeneity	3.41	2.28
16	Custom product	5.10	2.20
35	Customer need	8.63	2.15
43	Dominant competitor	4.89	2.09
37	Market size	6.53	2.09

FACTOR SENSITIVITIES & INDIVIDUAL EFFECTS*

FACTOR	MINUS* EFFECT	PLUS* EFFECT	INDIV* EFFECT
1 Product Superiority/Quality	68.6%	81.0%	63.3%
2 Economic Advtg to user	70.0%	79.9%	52.7%
3 Market Need, Growth & Size	65.8%	83.0%	58.6%
4 Market Competitiveness	78.7%	71.5%	55.4%
5 First into market	77.5%	72.8%	56.9%
6 Overall Company-project fit	70.3%	79.6%	65.1%
7 Technological synergy	72.9%	77.4%	50.1%
8 Market newness to company	76.0%	74.4%	58.4%
9 Project Determinateness	73.5%	76.9%	53.2%

*) NOTE: Minus and plus effects denote the probabilities of success if that factor were to be decreased respectively increased by one standard deviation. See factor impact table for means and deviations

Indiv Effects show the individual or 'stand alone' effects of each factor i.e. if all other factors were neutral or zero, what would then be probability of success.

DanProd Evaluation Work Stages

We recommend the five work stages below for the application of the DanProd Model. In sections 1-5 below, the details of the particular stages are reviewed.

1. Preparation for Evaluation Meeting:
The product concept is described using the lay-out of the "Concept Description"form. This form generates a consistent, adequate presentation of the product concept to the participants in the subsequent evaluation meeting. This form is filled in by the concept holder and/or the project manager.

2. Evaluation Meeting:
For a first-time application of the DanProd model, the participants will be given an introduction to the background and application of the model.

After this, each product concept is introduced to the participants, using the concept description as the starting point. This introduction is conducted by the concept holder or the project manager and will usually last for about 15 minutes, incl. the answering of questions made by evaluators.

Each evaluator will evaluate individually the qualities and chances of success of the concept using the 48 criteria integrated in the DanProd model.

3. Analysis of Product Concept:
By application of the DanProd model, an analysis is then made of the chances of success of the product concept, as well as its strong and weak points. The person in charge of the meeting will analyse the printout with a view to the preparation of the follow-up meeting.

4. Follow-Up Meeting:
The results relating to each particular product concept will be reviewed and discussed. Any causes of disagreement will be discussed, and the evaluators will, on the strength of this, make changes to their assessments if so required.

A plan of action is devised for the course of action relating to each particular product concept, i.e. whether the project should be stopped or continued, and which analyses should be made or action taken in order to reduce the amount of uncertainty of the particular project.

If more than one product concept is involved, the product concepts may also be given interrelated priority.

5. Final Processing and Conclusion

Based on the results of the follow-up meeting, a brief report is made out (one or two pages) and handed to the participants. In addition , the concept is subjected to further analysis by applying the DanProd model (any assessment corrections are entered through the keyboard).

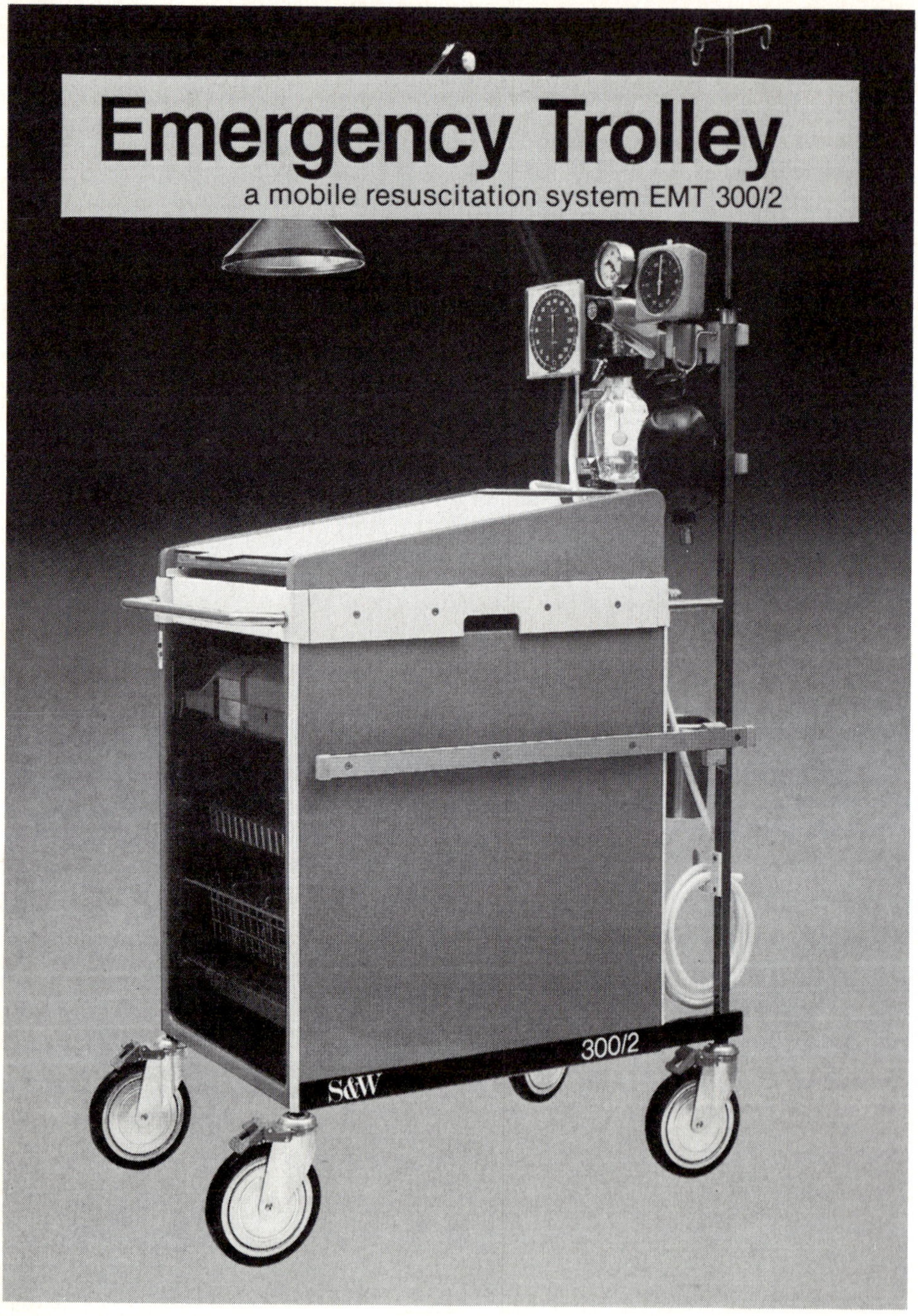
Emergency Trolley
a mobile resuscitation system EMT 300/2
300/2
S&W

The Rescar 3 Case

One of the products evaluated by S&W Medico Teknik using DanProd is the Rescar 3. Rescar 3 is a further development of an emergency trolley shown in the picture on the preceding page.

1. Description of Product Concept

An emergency trolley for revival purposes. The trolley contains all the equipment required to allow for revival using the "basic life support" and "advanced life support" techniques, incl. a monitor and a defibrillator.

The trolley is to be smaller, lighter, and preferably cheaper than the known EMT 300/2.

2. Description of Market, and the Products Use

The product is to be sold primarily in Far East markets and, to a lesser degree, in Denmark and Europe where the particular hospitals are inclined to make trolleys of their own. The clientele consists of hospitals. The product is to contain all relevant equipment for revival. It is to be built up the modular fashion so that particular components may be removed immediately and quickly be brought to the relevant place if you do not want to bring the whole trolley.

3. Technical details of the product

From a purely technical point of view, this product is simple. It consists of a steel pipe frame and a plastic material cabinet plus various drawers. An important unit which has to be available at the revival trolley is that of the defibrillator. A new and smaller type will have to be developed. The size and shape of this product is important to the design of the emergency trolley.

4. How is the product better than, or differs from other

Actually, there are no major competitors. The primary competitio relates to trolleys designed by the hospitals themselves. An important competition parameter is the choice of the proper trolley equipment. Further, the trolley is to distinguish itself in being smaller and more handy so that it may, for one thing, be easily transported up and down stairs.

5. Why should our company seriously consider this product?
From a strictly technological point of view, this product is fairly simple to produce. However, it has now become possible to make this trolley smaller since the equipment which is required to be available at the trolley has become in itself smaller, due to the technological development. By itself, the trolley will fetch a good sales price, and it will also entail sales of other products, including particularly defibrillators.

Result of DanProd Concept Evaluation of Rescar 3

The DanProd Rescar 3 printout is on some of the previous pages in this paper. The first page of the printout shows that the success probability rate has been computed at more than 70%. The product concept may thus be described as highly promising.

The strong and weak points of the Rescar 3 project would appear from the project profile on the first page of the printout (see page 0.5). This project is positive with regard to product superiority and quality. As regards user economy, it is marganally negative. Direct savings from using this product will in fact not constitute any major selection criterion for the client. It is far more important to which degree the product may perform some functions required for the customer to be performed in a manner fast and effective.

Factors 3 to 5 of the project profile describe what the product market situation is like. These factors are marginally positive.

Factors 6 to 8 of the project profile show how well this project is adapted to the resources and knowledge of the organization. The project is positive in respect of Factor 6 "Overall CompanyProject Fit", particularly with respect to marketing, sales and distribution channels, and overall management. Factor 7 shows that the product is not particularly well adapted to the organization with regard to the technology. This is due to the fact that this company is primarily an electronics manufacturer, less geared toward solving mechanical problems. Factor 8 is maginally positive.

The last factor, "9 Project Determinateness" illustrates how clearly the product concept has been described, and how clearly the technical solutions have been specified. On this score, this project is marginally negative.

Detailed Analysis of Answers to the Particular Questions

Generally, this project looks highly positive, and would seem not to be particularly hazardous. But how sure can you be of this conclusion?

In order to throw light on this, one may consider in which respects evaluators disagree, and also ponder how they themselves evaluate their certainty in answering the particular questions. These aspects may be read from the printouts.

It would appear that there are 20 questions causing much disagreement which may generally be said to be a rather large number. This project was one of the first against which S&W tested this model, and in such cases it frequently turns out that the disagreement is to some extent due to the questions not being perceived alike by all evaluators.

In spite of this, the following evaluation of the project did, however, show that on various questions there was in fact true disagreement about and uncertainty in making the answers. The disagreement on questions 6 and 3 thus manifested uncertainty as to whether they should manufacture the product themselves or whether it should be made by a sub-contractor. There was also an amount of disagreement as to whether the existing subscontractors were qualified to handle the organization of the production and the actual production. This was a point for further investigation in connection with the more detailed project planning.

Question 16, on the point of the product being especially adapted or not, also related to a problem to be dealt with further when preparing the concept; i.e. should a standard product be made, or should a concept be generated which lends itself well to being tailor-made to the needs of the particular customer.

Question 30 as to whether the product performs a unique task for the customer was also a question causing much disagreement. This discussion led to a more specific description of the product target group, and of which functions the product should have in order to solve and meet the problems and needs of this particular target group.

Questions 33 to 47 on the product market situation are generally marked by much disagreement and lower certainty values than are the other questions. And this in turn led to the recognition that the company knowledge on the market and the competition was generally not good enough. This may be viewed in the light of the fact that the product has not enjoyed the patronage of the company management to any market degree. In spite of this, it still has sold fairly well.

Project Rescar 3 Conclusion

The product should be developed within a period of one or two years. The development work should be coordinated with the development of a new defibrillator. When preparing the project specification to form the basis of the final decision on the fate of the project, particularly the questions below should be clarified:

- how and by whom should the product be produced
- a more specific description of the target group and its problems and needs
- a better description of the market situation and the competition

References

Jens H. Arleth, Robert G. Cooper, Henning C. Juhl, Vurdering af nye produktideer. (Assessment of new product ideas), Technological Institute, Taastrup, Denmark, 1986. ISBN 87-7511-653-7.

Robert G. Cooper, A guide to the evaluation of new industrial products for development. Industrial Innovation Centre, Montreal, Canada.

Innovation: Adaptation and Growth, edited by R. Rothwell and J. Bessant
Elsevier Science Publishers B.V., Amsterdam, 1987 — Printed in The Netherlands

A CASE STUDY OF PRODUCTION DESIGN BASED ON THE FUSION MODEL

N.HAYASHI[1], K.ISHII[2] and T.ICHIMURA[3]

1Department of R&D, Asahi Ohkuma Industrial Limited,Asahimae-cho 5050, Owari-asahi-shi, Aichi 488 (Japan)

2Department of Industrial Engineering, Kanazawa Institute of Technology, Ohgigaoka 7-1, Nonoichi, Ishikawa 921 (Japan)

3Department of Management, School of Commerce, Nihon University, Kinuta 5-2-1, Setagaya-ku, Tokyo 157 (Japan)

SUMMARY

This paper develops a method of an effective production design to harmonize quality,due date and cost of new/improving product, and presents the result of its application to a painting machine case.

The method proposed in this paper is based on "Fusion Concept","Product Characteristics Development" and "Product Family Tree".

The results allow us to identify the key point of process control for quality assurance program activities in manufacturing stage and the item of management by priority for the Total Quality Control.

INTRODUCTION

Almost innovation and improvement of product have been developed by a try and error steps of the persons in charge of research and development, and the evaluation of innovations is mainly focused on only the 'success or failure' of the results rather than the causes and process. So that, it is almost impossible to find out the clue to improve or redesign the management system for product innovation.

We have developed and proposed the models and methods to analyze "Information behaviour" in user's needs assessment and product innovation process to improve or redesign the product innovation management system. Those are, "Fusion Model", "Product Characteristics Development","Stimulus-response Model" and "System Module Analysis". The fusion model aims to clarify the function both of needs assessment and technology assessment and to fuse the both functions into innovative ideas(ref.1,2). The product characteristics development aims to clarify the relation among four product characteristics in the fusion process such as "merchandise characteristics","technical characteristics","manufacturing characteristics" and "marketing characteristics"(ref.2). The stimulus-response model aims to clarify the relation among the product characteristics, user's satisfaction and user's behaviour(ref.3,4). The system module analysis aims to clarify the relation among input, goal and output in the fusion process and to identify effective information sourse for "Idea Generation"(ref.5).

Although we have analyzed the information behaviour in product innovation process by the use of the models and methods mentioned above, an useful method to analyze the developing process from the technical characteristics into the manufacturing characteristics has not gotten. While quality assurance program activities in process control are successful in most Japanese industry, but the management systems for total quality control(ref.6) including user's needs assessment have not been developed sufficiently.

It is considered that one of causes of the above problem is an obscurity of the process in production design which connects user's needs with control characteristics of the manufacturing process. In this process, quality assurance-,cost reduction and maintaining due date should be harmonized. The method proposed in this paper is based on an idea combining the product characteristic development and "Product Family Tree" to solve the problem mentioned above.

The aims of this study are

1)to propose a method for process design/routing plan based on the fusion concept,

2)to apply the method to the case of painting machine and to discuss the results.

SOME ASSUMPTIONS

We have made some assumptions in this study.

A model of information behaviour

Information behaviour in the product characteristics development and the process control activities for quality assurance program based on the fusion concept is shown in Fig.1.

In this model,the merchandise characteristics are determined on the basis of informations concerning politics,society,economics and the surrounding conditions. The merchandise characteristics are the product characteristics required by users,which are represented in terms of user's languages. The merchandise characteristics are translated into the technical characteristics along with a target for development. The technical characteristics are the characteristics which represent the merchandise characteristics as engineering measurements such as "cm","Kg" and "Km/hr.". The target for development might be decided by a policy,on the basis of analysis of product line,competitive product,user's satisfaction and claims. The manufacturing characteristics are decided from the technical characteristics on the basis of the target for development. The manufacturing characteristics are the control characteristics of manufacturing stage, into which are translated from the technical characteristics. In this stage,many cost reduction programs might be planned and the key points of the process control activities for quality assurance program are decided. Based on

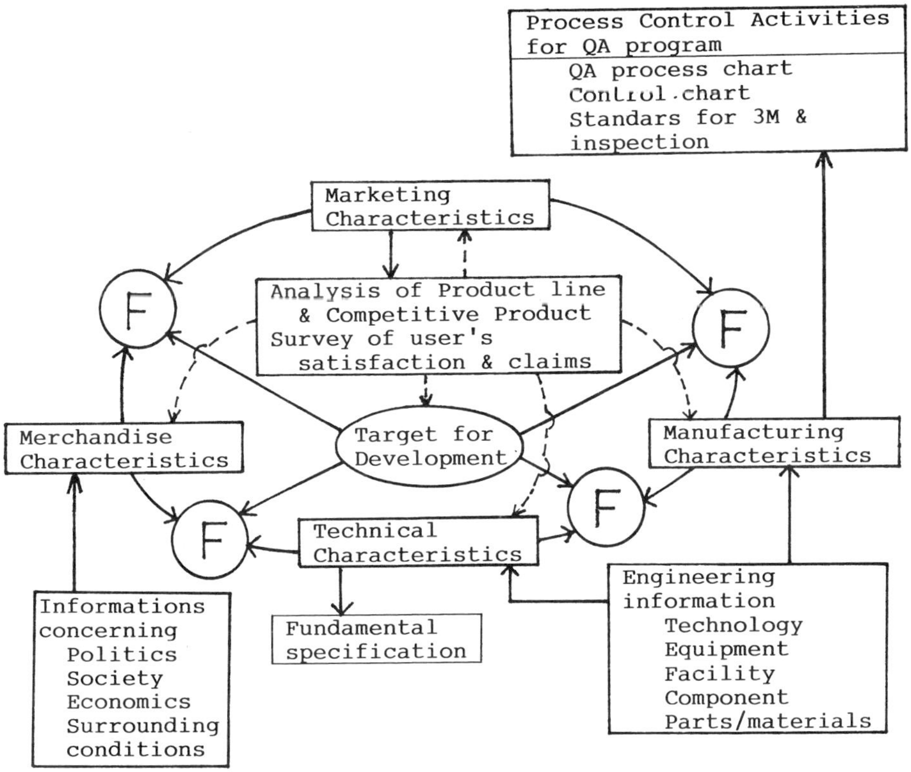

Fig.1 MODEL OF INFORMATION BEHAVIOUR BASED ON THE FUSION CONCEPT

Note; →(F)← Fusion, ⟶ Planning Information Flow, - - → Feedback Flow

the key points for the quality assurance program, management tools such as quality assurance process chart(QA process chart),control chart and some kind of standards are established and used to achieve effectively the daily program. The marketing characteristics are decided from the merchandise and manufacturing characteristics. The marketing characteristics are the characteristics which shoud be considered in selling,advertising,service and distibution activities.

The results of the survey of competitive product, user's satisfaction and claims with the product marketed might make change or reinforce the tagrget for development, four kind of product characteristics and the points of process control activities for quality assurance program.

Product family tree and family tree analysis

"Product family tree" is an useful method to represent the relation among product,components and parts by a tree type diagram considering "level" concept and it is called "Bill of Materials"(ref.7,8). In this study, it is called "Family Tree Analysis" to present the developing process from a fundamental functional specification of a product, which is decided on the basis of the technical characteristics(ref.5), into the manufacturing characteristics by using the product family tree.

The specification makes the fusion process develop into production design stage, in which many cost reduction ways are considered and tried under the restrictions of maintaining quality and due date, and the result of the stage is represented as a process/routing plan.

The procedures for the family tree analysis are as followings;

1) To present a fundamental functional specification as a product family tree.
2) To clarify the relation between the technical characteristics and the combinations of the components or parts of the product.
3) To clarify the relation among the fabrication/procurement ways such as by home production line, by subcontractor or purchase, and the components and parts from the points of a process capability, cost and due date,
4) To clarify the relation between the conditions of cost/due date, and production systems(Note 1).

Based on the process design/routing plan decided through the above analysis, the manufacturing characteristics are developed, and the process control characteristics for quality assurance program(QA) are established in each stage such as processing, inspection and storage.

Fig.2 shows an example of the family tree analysis in case of A product consisted of three components which have three parts. Fig.2(1) shows the original product family tree which is directly made of the fundamental specification of the product developed. One of the technical characteristics might be affected directly by both components A,1 and A,2. So that, an inspection in process might be needed immediately after assembling components A,1 and A,2. When the component A,3 is fabricated by a subcontractor, it might be neccessary to establish an inspection for acceptance. The sub-assembly component consisted of components A,1 and A,2 might be fabricated and stored in the storage area, if two kind of products A and B are produced intermediately in a final assembly line to maintain due date and increase utilization of the line. Through the above analysis, the final family tree in Fig.2(2) and the process chart of the final process design/routing plan in Fig.2(3) are obtained.

The method of product characteristics development

Information behaviour in the fusion process of product innovation has been assumed to be presented as the information development process for four kind of product characteristics such as merchandise, technical ,manufacturing and marketing characteristics. To analyse the information development process, the product characteristics sheet(ref.2) is used, and the concept of level in each set of characteristics is introduced to present the process of information development.

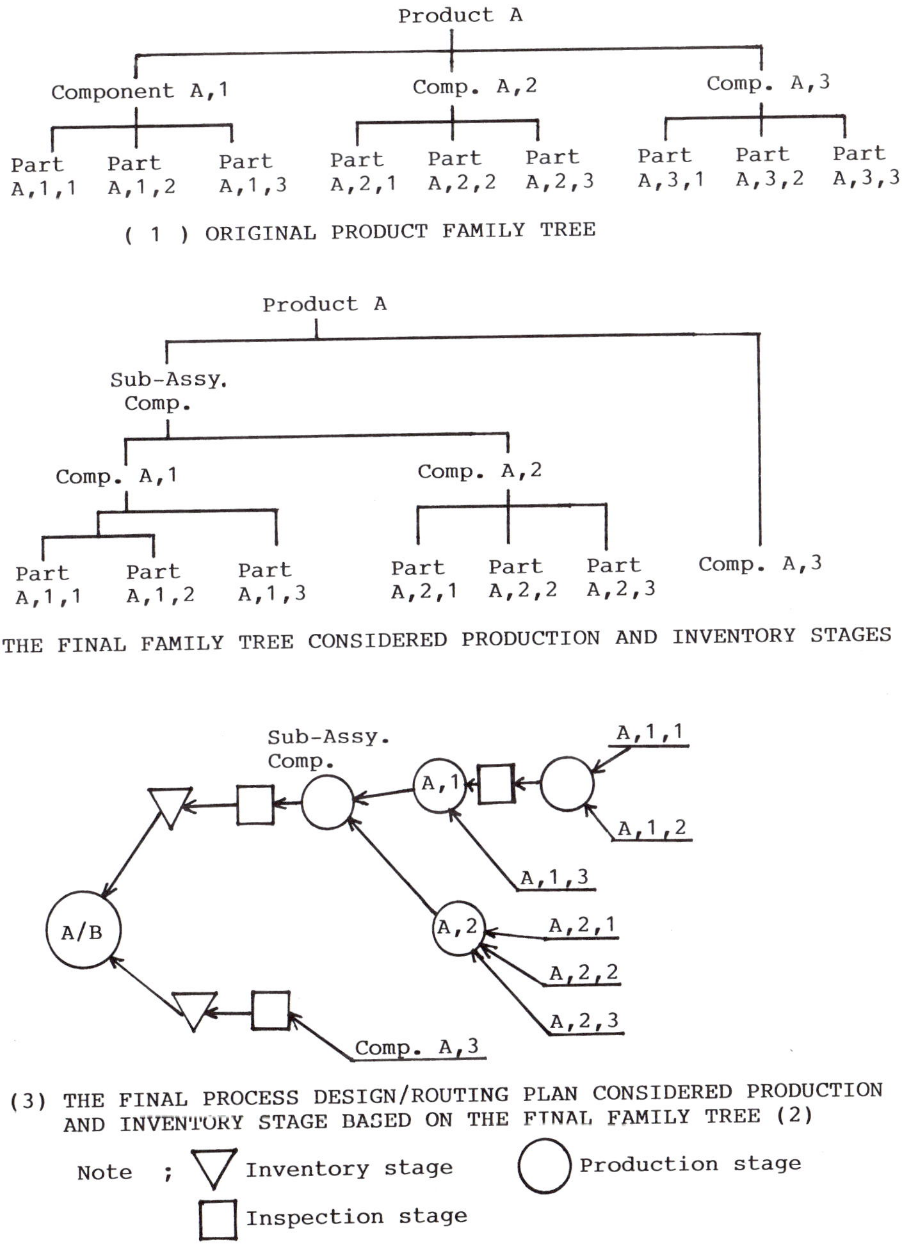

Fig.2 THE FAMILY TREES AND PROCESS CHART OF THE FAMILY TREE ANALYSIS

CASE STUDY

Background of the development and the target for the development(Note 2)

(i) Policy for the develoment. The company was founded in 1942. It has about 400 employees and sells about 8,500 million yen in 1986. The capital of the company is 255 million yen. Main product of this company are paint spray equipment and cold forging machine. The development is started under the following policies;

1) to develop new technology by the R&D department of the company,
2) to use the existing market channel of the company,
3) to add new sales point into those of existing and competitive products,
4) to improve the level of sales point of the existing product.

(ii) Analysis of product line. The paint spray equipment have some product groups according to the difference of painting method. Table 1 shows the result of analysis of paint spray equipment product groups. Based on the result, the company decided to increase the market share of the product group a.

Table 2 shows the situation of the product group a's market of Japan at the beginning of 1986. The company Y was almost monopolizer of the market, because the company Y had patented products in the market. After the patent right was free in 1981, the company X had participated the competition. The participation to the competition is so later than competitors Y and Z, that the market share of the company X was the smallest of the three. While the company X had shipped

Table 1 CLASSIFICATION OF THE PRODUCT GROUPS

Total Market \ Company		Market Share		
		Hight	Medium	Low
Rate of Growth	Hight	c		a
	Medium		d	
	Low	b		

Note ; The blank of the cell means no product group

Table 2 THE SITUATION OF THE MARKET OF PAINTING MACHINE

COMPANY SCALE \ INDUSTRY OF USERS	A	H	F	M
Large	Y	Y	Y	Y
Medium	Y	Y>Z>X	Y	Y>Z>X
Small	Y	Y>Z>X	Y>Z>X	Y>Z>X

Note ; "Y" in the cell means that the company Y is the monopolizer of the market
"Y>Z>X" in the cell means that the company Y has the largest share of the market and the company X has the smallest share

some products into the market, a little market share and some user's claims were obtained.

Considering the result of analysis of product line and the situation of the market, the company X started the development about the middle of 1986 with the following aims;

1) To focus on the market which company Y monopolizes,
2) To increase the market share by getting the company X's own market.

(iii) Target for development. Based on exhaustive analysis of competitive products and informations with user's claims, the targets for the development were decided as follows,

1) ability of painting various shape of the works with various level of painting quality,
2) to prevent production line from stopping caused by painting machine troubles,
3) ease of cleaning.

Results and discussions

The development is being progressed at the stage of mass production trial May in 1987.

Table 3 shows a part of the result in the information development from the merchandise characteristics into the technical characteristics and the manufacturing characteristics.

Fig.3 shows a part of the original family tree made of the fundamental specification of the painting machine.

Table 4 shows the relation between the technical characteristics and the parts/components of the product which are presented as a product family tree.

Fig. 4 shows a part of the final process design/routing plan.

The application of the family tree analysis and the method of product characteristics development to the painting machine case enables to clear the followings;

(1) The relations among the merchandise characteristics which are user's needs, the technical characteristics and the manufacturing characteristics can be made clear.
(2) 40 manufacturing characteristics can be decided according to the target for the development. These characteristics make it possible to decide the key point of the effective process control for quality assurance program.
(3) The family tree analysis is effective for quality assurance but is not so effective for cost reduction and maintaining due date in this case. The reasons are considered as follows;
   1)the restriction for due date is not so severe,
   2)the number of components/parts consisted of the product is not so sufficient

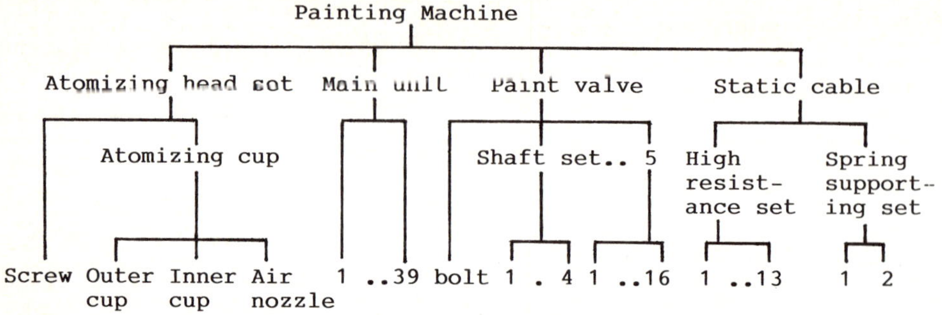

Fig.3 A PART OF THE ORIGINAL FAMILY TREE OF THE PAINTING MACHINE

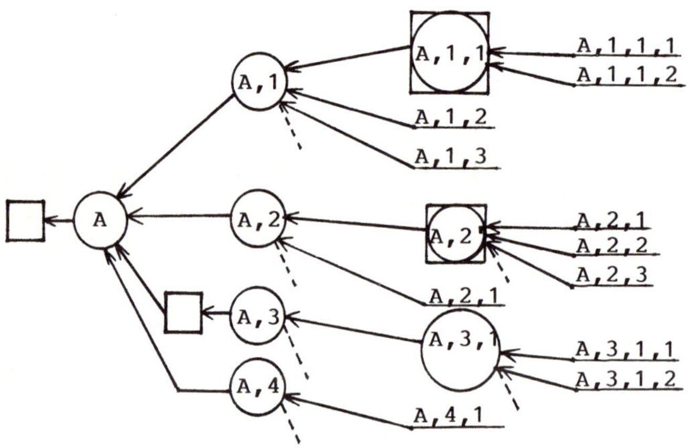

Fig. 4 A PART OF THE FINAL PROCESS DESIGN/ROUTING PLAN OF THE PAINTING MACHINE

as being able to create many alternative plans,

3)the fundamental specification is designed by using so many original parts, that few standard parts can be used.

(4) The family tree analysis is so effective to assembly processing but it is not so effective to machining/heat treatment processing. In order to reinforce the family tree analysis for the above disadvantage, it is an effective way to combine with a process analysis chart.

CONCLUSIONS

In this study, the basic method for effective production design to harmonize

Table 3 A PART OF THE RESULTS IN PRODUCT CHARACTERISTICS DEVELOPMENT

ELEMENT	MERCHANDISE CHARACTERISTICS		TECHNICAL CHARACTERISTICS				MANUFACTURING CHARACTERISICS
Effi-ciency	It can select spraying performance in accordance with figure of coating object	It can spray passing into concave part of coating object	Rate of penet-ration	Peripheral speed of cup edge	Cup diameter		Accuracy of atomizing cup diameter
					Rotational frequency		Inspection standard for
							00000000
							00000000
				Air as-sisting device for Air pattern	Air velocity	Air pressure	
						Air nozzle orifice dia.	00000000
			≈		Air volume	Air pressure	
						0000000	00000000
						No. of orifices	Accuracy of air nozzle distribution angle
					Air blow direc-tion	Jet angle of air nozzle	Accuracy of nozzle processing angle
		It can spray back side of coating object	Rate of wrap around ≈				
		There are few touch-up coating parts of	Rate of achievement ≈				

Table 4 A PART OF THE RELATION BETWEEN PARTS/COMP. AND THE TECHNICAL CHARACTERISTICS OF THE PAINTING MACHINE

Family Tree	Technical Characteristics
Painting machine A	Rotational frequency Weight Size
— Atomizing head set A,1	
—— Atomizing cup A,1,1	Unbalance rate
——— Outer cup A,1,1,1	Groove pitch 000000000 000000000 Shape of cup Cup diameter V-groove angle
——— Inner cup A,1,1,2	Liquid colliding angle
—— Air nozzle A,1,2	NO.s of orifices Jet angle of air nozzle Air nozzle orifice
— Main unit A,2	
—— Turbo set A,2,1	Unbalance rate
——— A,2,1,1	
— Paint valve A,3	Weight Size NO.1 item
—— A,3,1	
——— A,3,1,1	
— Stafic cable A,4	

quality, due date and cost of new/improving product has been developed on the basis of the fusion concept, product characteristics development and family tree. The followings are obtained for the usefulness and limitation of the proposed method;

1) It is possible to present the process of production design which is product characteristics development from the technical characteristics to the manufacturing characteristics.
2) The key points of process control activities for quality assurance program can be determined, in which the relation between user's needs and the process control characteristics are clarified.
3) It is neccessary to reinforce the family tree analysis in application to the processings except assembly processing.

In recent years, many companies in the world have made a great effort to achieve total quality management. For example, a CAD/CAM system is one of the useful tools for the activities by deriving benefits from innovation of computer technology. The CAD/CAM system aims at deciding effectively product and production plannings to harmonize quality, cost and due date by the use of computers. It is the most important point for developing an effective CAD/CAM system to clarify information behaviour from user's needs to process control characteristics.

Note 1

The quantitative analysis method for this problem has been proposed in the reference 9.

Note 2

The description of the contents is follows to the "Fusion model for developing new product" of the reference 2.

REFERENCES

1 K.Holt, Product Innovation, New-Butterworths Management Library, 1977, pp.61-89
2 T.Ichimura,R.Muramatsu and K.Ishii, A Method for analyzing information behaviour in needs assessment, in: H.Hübner (Ed.), Proc. Int. Conference on Product Innovation Management, Innsbruck, Austria, August 26-28, 1985, Elsevier, Amsterdam, 1986, pp.131-141.
3 K.Ishii and et al., Users' needs assessment by single and comparative assessment methods, in: H.Hübner (Ed.), Proc. Int. Conference of product Innovation Management, Innsbruck, Austria, August 26-28, 1985, Elsevier, Amsterdam, 1986, pp.143-153.
4 M.Yamashita, Y.Seki and T.Ichimura, Method of brand name designation by user's needs assessment, in: H.Hübner (Ed.), Proc. Int. Conference on Product Innovation Management,Innsbruck, Austria, August 26-28, 1985, Elsevier, Amsterdam, 1986, pp155-163.
5 Y.Neo and et al., Development of an automatic cleeaning machine; Case study

based on the fusion model,in: H.Hübner (Ed.), Proc. Int. Conference on Product Innovation Management, Innsbruck, Austria, August 26-28, 1985, Elsevier, Amsterdam, 1986, pp.165-176.

6 A.V.Feigenbaum, Quality control-Principles,Practice and Administration, Mc Graw-Hill, New York, 1951

7 T.Hirano, Research on the Standardization of Product Design -Especialy on the Standardization of the Hierarchy Parts Structure of Product-, Doctoral Thesis of Waseda University, Tokyo, 1983, p150 (Japanese)

8 R.Peterson and E.A.Silver, Decision Systems For Inventory Management And Production Planning, John Wiley & Sons, New York, 1979, pp.450-483

9 T.Tabe and R.Muramatsu, The economical Assessment and Design Method of Some Production Ordering Systems in a two-stage Production and Inventory Model,in-:D.M.Zelenović (Ed.), Proc. Int. Conference on Production Research, Vol.1 Novi Sad, Yugoslavia, August 24-29, 1981, University of Novi Sad, Novi Sad, 1981, pp.329-333

Innovation: Adaptation and Growth, edited by R. Rothwell and J. Bessant
Elsevier Science Publishers B.V., Amsterdam, 1987 — Printed in The Netherlands

PROCESS OF TECHNOLOGICAL INNOVATION AND THE NEW ROLE OF PRODUCTION IN STRATEGIC MANAGEMENT

Dr. CESAR CAMISON ZORNOZA, Escuela Universitaria de Ciencias Empresariales, Valencia, Spain

THE STRUCTURAL CHANGE IN INDUSTRIAL COUNTRIES

Since the middle of the 1950's decade in the United States and some years later in the rest of developed countries, a new glimpse can be observed in - the ensemble of inteconnected structural tranformations that have radically- modified the environment that created the model of mass production.

Among these meta-enterprise powers we can stand aout:

1) An increasing competitive offer proceeding both from new competitors in - the advanced countries themselves and from the new industrial countries. The increasing competition in most of the industrial sectors has been motivated by an excess of installed productive capacity, by the definition of economic growth and competition as primary managerials targets, which provokes- crossed inter and extra-sectorial attacks, and most of all, by the competi-- tion of new industrial countries that have joined increasing better cualities, service to the customer and distribution networks, to their custormary advantage in prices due to minor production costs.

2) A remarkable rise in workers training level, that renders anti-economic - keeping them on in semi-qualified jobs, according to taylorist industrial -- organization. New criteria are needed nowadays to involve workers in innovation and creativity, essential bases of productivity and quality.

3) The dramatic increase in information received by consumers about the in-- ternational offer of products, joined to the rise in level of life and sub-- sequent enrichment of cultural power, have provoked a permanent tendency to demand diversification and at the same time and increasing trend to products of quality.

The disparition of mass demand and consequent loss of market homogenity, both factors connected to the recent consumer's appreciation of the "quality of the difference" have slowly broken the prevailing "series economy" and - the mass production aiming to reach the greatest "cost economy, mainly for- two reasons:

a) The efforts of marketing, based on trade-mark, advertising and sales - promotion, have only managed to give the consumer a partial choice allowing him to put forth his individuality, "underlining the non-functional factors that go with a product. But the personal bias is more complex, the marketing being unable by itself of offering the real variety the customers demand.

Consequently, the new environment coordinates demanded a "demassification of the offer" greater than the one sponsored by the mass production model -- with marketing efforts.

b) In order to put into practice the "multiple links technology" (ref. 1) a-standard product in repetitive form is required and the maintenance of a -- constant production volume, to standardize the consumed inputs proportions - too. These requirements of the rigid technologies concomitant with the inter-changeable production have proved incompatible with the "demand demassification", and unable to conjugate the costs reduction with the increasing produced at the same time of the response to the changes of environment and to face up its new pressures: shorter delivery time; reduction of the product life - cycle and more frequent modifications; better quality demand, both by the -- marketing department that wants to base the mark image on the service record of his products, and by more strict specifications proceeding from the buyer and by the normalization; special series etc.

These new vectors of the managerial environment have cause the entry of - the production variety cycle in a new phase, characterized by the descent of mass production. This production model, in spite of its being kept as a ma-- nufacturing model, prevailing to 1970ś decade, is now being questioned and - replaced by new production concepts, more adjusted to the background struc-- tures. Accordingly an increasing importance of the discrete production in -- reduced lots or batch production can be seen ererywhere.

This transformation of the industrial economies has been based on a great number of process technological innovations, PTI, so as Electronics, Infor-- matics and robotics. The revolutionary techological change gained by the mi-croelectronics progress had feasible a radical transformation of productive structures, thanks to the implementation of a myriad of TPI, generically na-med "flexible automation systems" (FAS". New principles of Resource Manage-- ment 50 was born on its influence.

"ECONOMIES OF SCALE" VERSUS "ECONOMIES OF VARIETY"

The mass production model was managed on the concept that skinner (ref. 2) named "production technically orientated", which it distinguish because top management forgot at the production function delegated on staff o short hie-rarchical level in manufacturing area, a great number of basic decisions re-lative to productive politics, The manufacture was dominated first by experts in efficiency (industrial engineers and informatics and cybernetics specia-- lists).

This resource Management conception, as confirms Starr (ref. 3), was the last phase of complete cycle strategic decisions. At the beginning, because economic progress basis es the technological dominion on a process, the production occupied a central place in enterprise.When this take off stage was surpassed, the Industrial Engineening advances permitted to diminish the variability and the production decisions risk likewise. Since then production was characterized as "reliable" because the manufacture options was smiles- and of prediction easy. When this decisions give up looking at outical, but no vital, decreasing to production by problems, the manufacturing questions was forgotten for enterprise top levels. The principal factors of the top direction decisions in this phase was finance: marketing and Research-Development.

Therefore, the concept of "production technically orientated" has provoked a cult of the efficiency in Industrial Engineening. The technique education of production executives and their isolation of top management has originated a Resource Management concept wich it predominantly es centred on daily inmediate problems in search of short-term benefits. This fixing on short-term has provoked a cost orientation deformed into system, which fost think that the production principal mission is to catch a good conbination of short costs, top quality and accpptable service to customers. This approach hides an implicit hypotesis: a plant with modern equipments, present methods, cooperative work force, a computrized information system, and a direction engineerelly instruct, has to cause an efficient direction of production. The "economy of scale" defence, which it was got with the process more and more intensive in capital, was hard up outcome.

This concept of production direction and the managers training more and more remote from technique disciplines had validity for decisions as: purchase of isolated machines for to solve local production problems ; or in productive systems very repetitive and with a traditional stagnant technology. Conditions was given during the splendour of mass production model. Because this is totally insuitable when we speak of TPI with basis on computer.

The new capitl goods and process planning techniques are derived from FAS and implant a direct defiance to all assumptions that "economies of sale" are based. most of all the idea that larger production volumens always shall generalt minor unitary costs. As contrasted with this opinion, the new technological capacities are established on "economies of variety" and aren't on the size of lot (ref. 4).

The characteristic main of the FAS, that it simultaneously is their essential advantage, rests on their capacity for to combine the starts of automatic mass production (economy, productivity, specialization and production rapidity), with the inherent positive points to the manual work (productive flexibility, top occupation of productive capacity, savings in capital investment and in industrial plant size, universality).

Therefor, the flexibility is conceived as the technologic-organizative -- support that permit to give an adequate answer to cualitative and cuantitative changes demanded by market at competitive costs" (ref. 5).

Davis (ref. 6) calls to new productive resultant model "production based on the client at mass production prices". Toffler (ref. 7) confirm that this model involves the restoration of preceding production system to Industrial Revolution, the craftsmanship, but it's moulded with a new technique basis of top technology now.

The implicit advantages group to TPI has procoked the idea that all productivity, new costs, quality, inventory level, or delivery time problem, - which it is rectified with integrated system developments based on new process technologies. But a great number of enterprises that have installed - FAS have verified as this technological innovations don't implicated improvements on conventional systems, or they don't generate a performance in -- accordance with their cost. In line with some estimations (ref. 8 and 9),- the failure rates can be placed around 50 %. The assymetric distribution of FAS are still to be explained now. Japan, West Germany or Sweden have advanced more than limited states in productive process integration with microelectronic technological innovations. In this way, whereas in 1930 all american industry has implanted near 30 FAS, the japonese firm Toyoda Machine Tool Co. only has very much (ref. 10)

These statistics have created a lo of questions. Whay do some FAS aplications fail, whereas others similar applications are successful in other en-- terprises? Why do some firms can introduce robotics systems with work force support, whereas others companies have to fight a battle for the same purpose?What do factors determine the speed and success in the fiffusion of - flexible automation innovations?

In spite of what es recent the studied phenomenon, japanes success and their growing industrial leadership analysis, so as very mech flexible automation projects, seem to prove that the decision only to adapt the technology more advanced don't takes to the success in automatic and necessary manner. The failure main cause in programs of FAS introduction and the advance by Japan or Germany or E.E.U.U. in production flexible automation, - rest on the american omission of special technology incorporated in TPI, which it do them don't to shave none of inherent tipical assumptions in manufacturing traditional equipment (ref. 11). The FAS applications that respect the assumptions of "technically orientated production" has reinforced the instrumental character of the technological innovations, which they are esteemed as sustitution processes of equipment and instalations - instead of as new experiences with technologies that open important consequences for the enterprise by their strategic dimension.

THE LINKS AMONG PRODUCTION DECISIONS, TECHNOLOGY, COMPETITIVE STRATEGY AND ORGANIZATION.

The advantage whole of the conjunction productivity-flexibility shows - repetible the necessity to corner the assumptions of the production traditional logic and to raise again the concept of "technically orientated production". This is specialyy sure when the mass production hid a capital fact: The Tripod technological option-manufacturing decisions - competitive strategie is interlaced for multiple bidirectional bows.

When the top management understand a FAS as a production means meeting - without implication on the production and enterprise strategies and organization, don't take on her analysis functions between the organizative and strategic connections and systems and process planning. That forget the important risks of thes conduct for the company, at once for motivation conflicts, now for rigidities at decisions adoption and responsability imputations, for pay systems, etc. The great flexibility and integrated nature of the TPAI paradoxically can impose a tyranny kind because they displace to-- others productive goods perhaps more adequate at the company strategic project, or they can obstruct the subcontracting.

But the influence inversely works too. The strategic imperatives don't only have take into account when it is bought a production system or they are adopted productive decisions. It's necessary to bear in mind the effect of available technological systems on the imaginable strategic options also. For example, the disponibility by Boeing of a CAM system has permited the - control of 1500 piece millions, which it's rather hand without the computers; a CAD system simultaneously has allowed to design very much plane complex - models. The product-market strategy of Boeing has been strongly influenced by the CAD/CAM technologies (ref. 12).

Sometimes, the possession of a CAD/CAM system can be a critical factor for to get a customer contract, who he have a compatible system with the first too, and he wishes to eliminate the problems of manual information transfers.

The great importance of these innovations is that they have opened new - competition lines based on the use of production as a competitive weapon. Therefore, the production role ignorance as strategic decision to take to important risks of competitivity losses given that the levels in quality, costs, productivity, delivery time, innovation or flexibility of FAS are aout of enterprise reach equiped with traditional technologies.

This new reality es specialy virulent for nature industries which are - situated in industrial nations, for some continuous flow industries, and - for the batch or individualized production enterprises which they are inserted into technology top sectors.

The technological basis of the standardized production industries was - founded on assebly lines, and it's was consistent with a competitive option centred about the price, in search of minor costs with a sufficient quality and a thin product differentation. The changes into offer and demand structure from the last decade have cause a conmotion in this competence framework.

The efficiency definition (as synonymous with low costs) as central sole objective of the production strategy was put under a ban then. There are - sufficient causes: an explosion of competitive products was born in new industrialized countries ; the diffusion of many TPI which they open the ways to alternative production models ; and the demand progressive of massification. The nature industries have to understand that the automation with traditional technologies or with flexible automation islands into a factory managed on mass production principles, are incompatible options owing to the - displacement to the unmassiveness without loss of continuity, now. It's equally clear that this phase coudn't to be caught with intensive process in work force, because this was precisely its main comparative disadvantage. The flexible integrated automation is the sole option to implant continuous productive systems but in reduced lots. The overlapping of TPI with microelectronic basis into these nature industries has thrown them to the top technology sector with more competitive force.

Even more, the manufacturing based on FAS is shaped as key element of the new entrepreneurs, who they will can compete in distinct markets and sectors perhaps the productive equipment ambivalence and the accomodation capacity of their operations management process and systems.

This new approach to the production direction will demand to the top management to retain some essential ideas for to face up to inversion decisions at FAS:

1.- to take into account that the FAS have important repercussions on the productive process, the productive system strategy, organization and nature, the production design and planning, the structure of the economic units, the labour conditions, and the demanded model of worker and executive, including the concentration, growth and location strategies, as soon as at the relations with the auxiliary industries. These transformations try to keep the - concilation advantages between automation and versatility which are propitiated for the productive flexibility strategy.

2.- The production function get in the strategic management as a critical factor, integrating the strategic development, the enterprise organization and the selection of process innovation projects a sole interrelated process which the top management and sistems and function experts both have to take part.

3.- The top management need to understand that she is the sole level with a capacity of the long benefits and strategic implications analysis which - they are derivated of productive area decisions. Therefore, she totally can't delegate on technician experts the manufacture management, the productive - system design and implement, and the evaluation of investment projects in FAS. The top management must impose the objectives and principles that guide the manufacturing activity, after to detect and to assume the kade-off which the FAS bears on the firm. This new approach have to permit ot the executive to understand and to manage the productive function.

4.- In order that it's possible, it's necessary to count on some expert advanced technology into top management, who he advise on the production - importance ento enterprise competitive strategies. It's often essential a "process champion" (ref. 13), a person delivered to the task to convince to the top management of the necessity for the new production concept. He have to be capable of resisting the modernization and technological change that inevitably arise into top management, now for little experience on that -- process, now for fear of the automation modify the function nature or the role distribution, now for fear of functional are divisional executives to lose autonomy owing to the productive process integrated character. Gunn (ref. 14) think that thes person have to occupy a Production Systems Development Vice-President into the staff, with the mission to guide and to manage the corporation to the Future Factory.

5.- As a consequence of these news organizative structures, the implantation of FAS will force to renew the concepts and methods that they have used in the product design and the design productive process coordination (ref. 15 and 16).

Moreover, the manufacture activities will have to adapt oneself to new ideas on production direction, which will shape the competitive picture of tomorrow leaders enterprises.

REFERENCES

1 Thompson, James D. ; "Organisations in action. Social Science bases of administrative Theory" ; Mc Graw-Hill ; New York ; 1967 ; pp 15.

2 Skinner, Wrckham ; "Manufacturing-Missing Link in Corporate Strategy" ; Harvard Business Review ; may-june 1969 ; pp. 136-142.

3 Starr, Martin K, ; "Modular production: a new concept" ; Harvard Business Review ; november-december 1965 ; pp. 131-142.

4 Goddhar, Joel D. y Jelinek, Mariann ; "Economías de la variedad basadas en la nueva tecnología" ; Harvard-Deusto Business Review ; second quarter 1985 ; pp. 73.

5 Quevedo, Manuel ; "Robótica y talleres flexibles. Ejemplos concretos de aplicación en empresas del grupo Ularco" ; paper at "First Forum on - Innovation and Enterprise" ; ASOCIACION para el PROGRESO de la DIRECCION ; Madrid ; 11 oct. 1983.

6 Macrae, Norman ; "The Coming Entrepreneurral Revolution : A Survey" ; The Economist ; 25 dec. 1976, nº 6956 ; pp. 55.

7 Toffler, Alvin ; "La tercera ola" ; Plaza y Janés ; Barcelona ; 1982 ; pp. 218-219.

8 Regentz, M.D. y Shaw, R. J. ; "How to Prepare users for a new System" ; Management Focus ; Peat, Marwick and Mitchell ; mars-april 1980 ; pp. 33.

9 Scott, Peter B. ; "The Robotics Revolution. The Complete Guide for Managers and Engineers" ; Basil Blackwell Publisher ; Oxfor 1984 ; pp. 218-220.

10 Bylinsky, Gene ; "The Race to the automatic factory" ; Fortune ; 21 feb. 1983 ; pp. 58.

11 Gold, Bela ; "La fabricación con ayuda de ordenador induce nuevos criterios de producción" ; Harvard-Deusto Business Review ; third quarter 1983 pp. 45.

12 Goldhar, J.D. y Jelinek, M. ; O. C. ; pp. 75.

13 Chakrabarti, Alok K. ; "The Role of Champion in Product Innovation" ; California Management Review ; winter 1974 ; pp. 58.

14 Gunn, Thomas G. ; "Computer Applications in Manufacturing" ; Industrial Press ; New York 1981 ; pp. 181.

15 Fisher, Anne B. ; "Behind the type at GM'S Saturn" ; Fortune ; 11. nov. 1985 ; pp. 34-42.

16 Alique, Manuel ; "Visión general sobre la situación actual en el mundo del tema robótico y talleres flexibles y su incidencia en el presente y futuro de la empresa ; paper to "First Forum on Innovationa...", o.c. ; pp.91.

Innovation: Adaptation and Growth, edited by R. Rothwell and J. Bessant
Elsevier Science Publishers B.V., Amsterdam, 1987 — Printed in The Netherlands

SUCCESSFUL MANAGEMENT OF THE INTRODUCTION OF NEW TECHNOLOGY

I. BARCLAY AND P.J. LUNT

Department of Industrial Studies, University of Liverpool,

P.O. Box 147, Liverpool L69. 3BX.

SUMMARY

Strategic corporate planning is the management process by which product manufacturing companies define how to adapt to change and remain successful. The development and application of an appropriate manufacturing strategy is fundamentally important to the achievement of the overall corporate plan.

The manufacturing strategy usually contains a major innovation which will involve substantial capital investment requiring detailed appraisal and justification.

We propose that current appraisal systems are mainly based on evaluation of the product/economic aspects. This appraisal tends to omit consideration of the actual operation of the newly installed equipment.

We will show quite clearly that current appraisal systems are mainly based on evaluation, and also, that a comprehensive appraisal system including both evaluation and operation will produce much greater benefits.

A methodology will be suggested as to how such a comprehensive appraisal system ought to be conducted. Fundamental in this methodology will be the attention paid to the management, organisation and integration of the proposed innovation.

INTRODUCTION

The nature of technological change has altered significantly with the introduction of computer based systems. Prior to these systems being used, the change in technology was limited in scope to the specific technology being used within a function, and was usually within the technological capability of the management of that function.

With the introduction of advanced manufacturing technology such as we have today, much more is required of the management function, and hence this aspect needs to be considered during the appraisal of the technology.

The aim of the research project on which this paper is based is to produce a comprehensive appraisal system for the introduction of new technology, which will encompass the areas of finance, technology and management.

The results of a questionnaire to support this research are currently being analysed. The aims of the questionnaire are twofold -

1. To investigate the effect of the introduction of new manufacturing technology on the management and organisation of the company.

2. To investigate the appraisal systems used for the introduction of the new technology.

The paper will aim to show that current appraisals of the introduction of new technology concentrate mainly on the "evaluation" aspects of the system, i.e. the financial and technological aspects, and tend to either pay little attention to, or ignore completely the implementation/operational aspects, i.e. the "managerial" functions.

We will also show that by using simple concept models, and by taking account of and applying previous proven work on creating a climate for innovation, a methodology can be developed that will improve the chances of success when introducing new manufacturing systems.

The paper is split into four main sections which cover:

I. The Problems Associated with Conducting Appraisals for New Technology.

II. The Changing Nature of Technology.

III. The Changing Role of Management.

IV. The Total Appraisal.

THE PROBLEMS OF APPRAISAL

Our research has concentrated mainly on the introduction of Flexible Manufacturing Systems into a company's current manufacturing operation. After visiting selected companies, interviewing senior management within those companies, and reviewing relevant literature, the list of factors used in the conventional appraisal of the proposed introduction of advanced manufacturing technology was produced and is shown below. (Fig. 1.)

1. Product Performance/Quality
2. Current Technology
3. Economics - External
   - Internal
4. Product Type and Quality
5. Corporate Strategy
6. Fear of "Being Left Out"
7. Desire to move forward
8. Technical capability of employees

Fig.1: List of Factors used in the Appraisal of Advanced Manufacturing Technology.

This work confirmed our view that conventional appraisal systems are based mainly on the short term economic and technical evaluation of the technology, and largely ignore the subsequent implementation, integration and operation of the new technology into the manufacturing operation.

When planning to introduce advanced manufacturing technology into a current manufacturing operation, the firm's management are faced with an almost bewildering set of inter-related problems, e.g.

- Which equipment should we buy?
- Will is be viable for our current <u>and</u> future products?
- How can we finance it?
- Have we got the technological capability?
- Will the workforce accept the change?
- How should it be managed and integrated?

etc.

The list of questions is lengthy, but ideally, all these factors need to be taken into consideration during the appraisal of new technology.

A typical time scale for the complete introduction of such a system is between 18 months and 3 years, and thus long term strategic planning is not really of practical help in answering detailed problems that might arise in the initial stages.

A Model for the Implementation of New Technology

Hutchin (ref.1) has proposed an "Implementation model" which, coupled with the economic/product evaluation aspects described previously, can produce a model for the total appraisal of new technology.

This model is outlined below (fig. 2):

Evaluation Phase
↓
Decision Making Point
↓
Implementation Phase
↓
Consolidation Phase
↓
Expansion Phase

Fig.2: Factors for Consideration in a Total Appraisal.

The relative "certainty" of the Evaluation Phase is counterbalanced by the relative "uncertainty" (at the start) of the Consolidation/Expansion Phase.

This feeling of uncertainty can lead to resistance to change by all levels of employees, but in particular, to middle management, whose jobs are most threatened by the introduction of advanced manufacturing technology (ref.2).

Whilst Hutchin's model is simple in concept, in reality it is complex, mainly due to the fact that the time scale from beginning to end is usually measured in years, (if indeed a defined end point is ever reached). Given the time scale, it is perhaps understandable that the concentration of management effort is placed on the short term phase (i.e. the evaluation).

There is also the fundamental problem that the evaluation phase is mainly the domain of the accountant, whilst the implementation and consolidation phases are the domain of the engineering/production function. It is not only their criteria that are vastly different, but also they use languages that are virtually mutually exclusive.

A MODEL FOR TECHNOLOGICAL CHANGE

The concept of introducing new technology can be represented very simply using an adaptation of the Sigmoid diffusion curve (Ref.3)

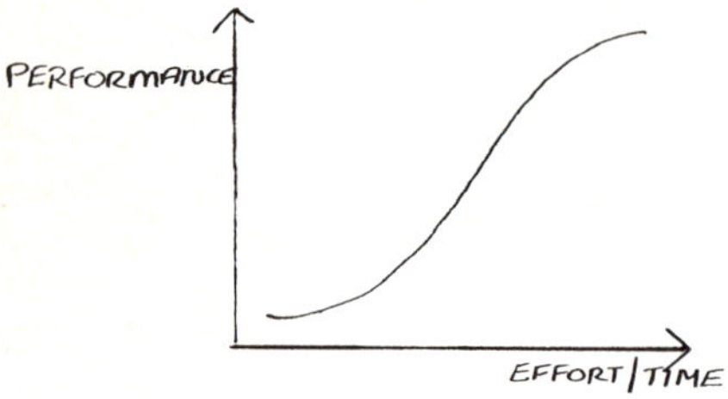

Fig. 3. The S - Curve

If we consider the diagram, we can see that after the introduction of the new technology, more effort put into increasing the efficiency of the technology results in an increase in its overall performance, until the stage is reached where considerably more effort is required to increase the overall performance slightly, i.e. the performance limit of the system is being approached.

Foster (Ref.4) made the point that:

> "Behind conventional management wisdom is the implicit assumption that the more effort put in, the more progress results."

This can be seen to be true for only the first half of the S-Curve. Thus, ultimately, a newer, more efficient system has to be introduced.

When a new technology is introduced to raise the performance of the manufacturing system, a "discontinuity" occurs, resulting in difficulties for the management and the workforce. This is shown diagramatically in Fig. 4:

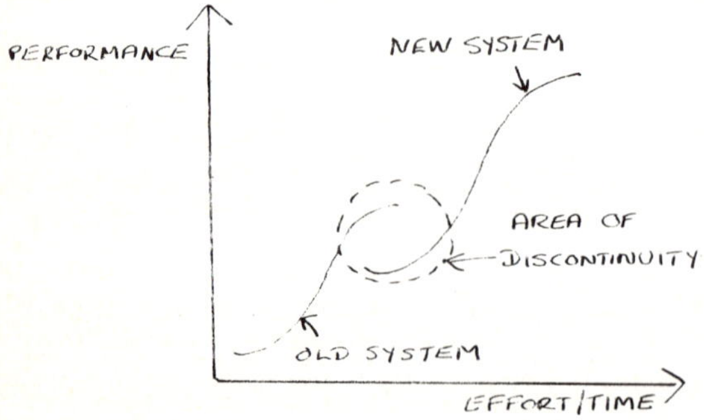

Fig. 4: Discontinuity and the S - Curves.

In the days of limited technological change, this discontinuity was small and could be easily controlled. With the introduction of advanced manufacturing technology, the discontinuity could be as great as the complete new system not being used at all.

"A system will fall into disrepute, i.e. it doesn't meet required performance, through either -

(a) Mis-use (c) Mis-understanding

(b) Dis-use (d) Mis-management

or a combination of all four." (Hutchin (Ref.1)).

"The roots of these are to be found at the decision making phase due to a failure to consider any/all of the following points:

1. Full problem identification
2. Full involvement of the workforce
3. Full appraisal of the system in context of the Company Objectives, the Business Plan, and the setting of system objectives."

Most companies compound the problem by not keeping records of technological productivity (Foster (Ref.4)), so they are not even sure of where their system is on the S - Curve. Thus, any management contemplating the introduction of a new manufacturing system is faced with a complex appraisal of inter-related problems ranging from defining the actual system to be purchased, through to the integration of that system into the overall operation.

THE MANAGEMENT FUNCTION

Let us now discuss the changing role of the "managers" of the new technology.

The Changing Nature of the Management Role

The complexity and interdisciplinary nature of the new manufacturing technology presents significant problems for the new type of "general managers" operating the new system. This "new" management role can be compared to the "old" role as outlined below (Fig.5):

Past	Current
Impact of technological change limited in scope.	Impact crosses functional lines
Technology changes technology-based industries	Technology changes every industry
Responsibility for technology delegated to functional managers	General managers take charge

Fig. 5. The Changing Nature of Technology Change.

These points are summed up by Marlow (Ref.5):

"Individual relationships and organisational structures are highly interactive - individual tasks are in line with overall project objectives; work groups are autonomous. There is a loosening of functional control."

Parkinson and Avlonitis (Ref.6) summarised the requirements of today's engineering managers in their paper "Management Attitudes to FMS", which provides insights as to what needs to be done to increase the adoption rate of FMS in British manufacturing industry. Their findings are grouped under five main headings, as listed below:

1. Technically-Orientated and Informed Management

User companies should identify knowledgeable technical people and provide them with the time and resources to evaluate the applicability of FMS technology in their own manufacturing operations.

2. Good Internal and External Communications

User companies should develop close and continuing contacts with the various sources of assistance of various kinds mentioned above, namely machine tool suppliers, research institutions and Universities, and Government.

3. Greater consideration of Training Policy

User companies should develop training policies to ensure adequately skilled and competent manpower and to create an infrastructure that will be ready to support FMS as it is installed in stages.

4. Greater Consideration of Long Term Strategy

User companies should develop a portfolio of innovative activity which includes strategic long-term thinking and action as well as short term innovation. Companies which are engaged in strategic planning are in a much better position to appreciate the contributions of FMS to the fulfilment of their business objectives.

5. Higher Levels of Risk Taking

User companies should learn to appreciate non-financial criteria in evaluating FMS technology, namely, the numerous "intangible" benefits flowing from flexibility. Relying heavily on analytical financial techniques biases investment decisions in favour of short-term innovations, and, as a result, major innovations, such as FMS, which promote long term technical superiority and competitiveness are not likely to be adopted.

All five are key management responsibilities which will be discussed later in the paper.

THE TOTAL APPRAISAL

Components of a Total Appraisal

Adair (Ref.7) proposed a model showing the inter-relationships between task, group and individual needs in a business environment. This model has been adapted to describe the scope of a total appraisal system as shown in the diagram (fig. 6)

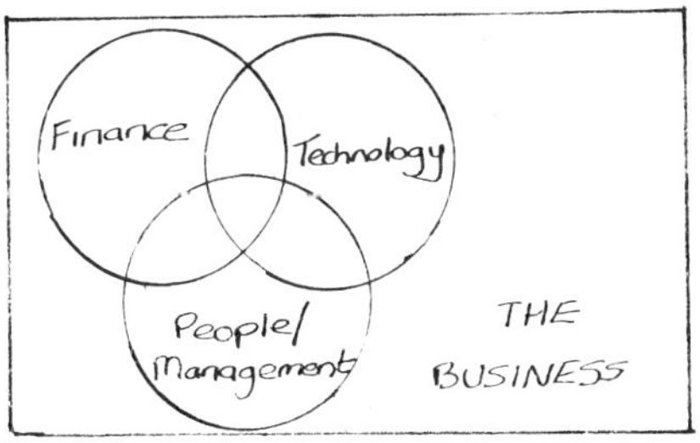

Fig.6: Model of a Manufacturing Company

It is worth spending some time briefly discussing each of these areas.

(i) The Business:

New technology is introduced to improve current product manufacture and accommodate the manufacture of future products. The new products and markets are usually defined in a strategic corporate plan. However, strategic manufacturing plans are less commonly found. Understanding your present technology will neither forecast nor accommodate any discontinuities due to the introduction of new technology.

To quote Foster (Ref.4): "Since most companies don't study S-curves, technology limits, or performance parameters, they don't understand where their technology can go, or what it will take financially to get it there."

Thus, companies need a strategic, manufacturing technology vision.

(ii) Technology:

The choice of technology is limited by the system manufacturers and the products made by the company. Hence, it is relatively easy to decide on the system required; therefore, the evaluation and decision phases can be executed efficiently. It is the integration of the system that causes most of the problems - this is mainly a management function which will be discussed later.

(iii) Finance:

The financial evaluation of new systems can be relatively simple (such as "gut feelings") or can take advantage of sophisticated computer packages such as IVAN.

The IVAN computer package is based on the work of Primrose and Leonard (Ref.8) and allows "the intangible benefits of introducing new technology to be quantified." The package uses an extremely detailed analysis based on traditional discounted cash flow techniques.

A student in the Engineering Faculty at the University of Liverpool, Rusooki, (Ref.9) has spent some time analysing the package and its use by companies who have purchased it. It was found that of many companies who purchased the package, the production directors in 20 of these companies did not know of the existence of the package, or who was using it within the company. The package was specifically designed for use by the production engineering team to aid the decision making process for the introduction of new technology.

The problem companies encounter with this package is that the required data was too detailed to be readily available, and so many subjective figures are used, thus taking away part of the credibility of the claim that the package quantifies the 'intangible' benefits of introducing advanced manufacturing technology.

In terms of measuring technological productivity, we have only come across one company from the ones we have visited to date, who carries out a detailed post-investment analysis to compare the actual and targeted performances of new technology. This evidence confirms Foster's comments (Ref.4) on technological productivity.

(iv) People and Management:

Considerable sums of money are currently being invested into the research of the management of new technology, this research project being one of many such projects.

Various groups of people have been investigating this field, including the Work Research Unit at ACAS, who have recently published a booklet (Ref.10) covering the topics of how to create the appropriate climate for change, the process of introducing change, the effects of the change process, and how to increase the chances of success.

Training of the management and workforce for the introduction of new technology usually receives some consideration within most companies, although we have not yet found evidence of a procedure for formally identifying training needs, and methods for implementing them.

The total appraisal for the introduction of new technology should take all the above-mentioned factors into consideration. A report has been published recently by Bessant and Haywood (Ref.2) which considers the total process of introducing new technology and in particular, flexible manufacturing systems.

The main conclusion from this report is that notenough consideration is at present being given to the <u>integration</u> of new technology in the workplace.

Theoretical Model of a Total Appraisal

In our previous model of a manufacturing company, we proposed three basic factors that make up the total business, i.e. Finance, Technology, and People and Management. At the appraisal stage, the prime consideration should be the finance/technology area, but some consideration should also be given to the people and management area.

However, because of the relative 'uncertainty' of this latter area (c.f. Hutchin (Ref.1)), detailed appraisal is neither feasible nor justified. What is needed is a different approach to the problem.

We are proposing than an organisational climate should be created which allows the changing needs of the people/management area to be activated and developed as the implementation and integration phases progress (c.f. Hutchin's Model of Implementation). This "climate of change" is an integral part of implementing the manufacturing strategy.

After the decision point has been reached, the areas of finance and technology decline in emphasis, and the area of people and management increases. This can be shown diagrammatically, as follows (Fig. 7):

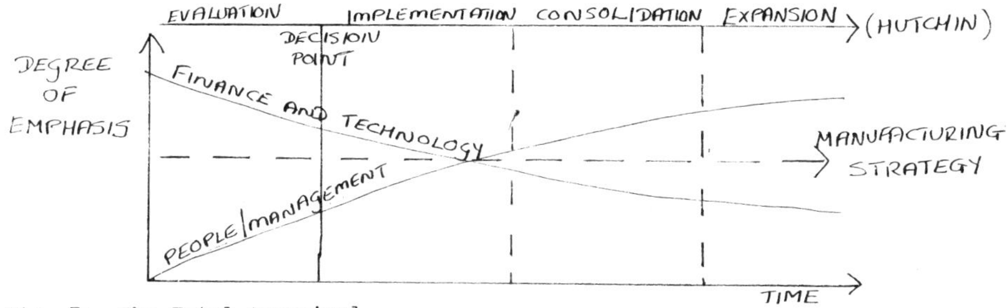

Fig. 7: The Total Appraisal

Various authors over the last twenty years or so have been investigating different aspects of this area and some of the main points are outlined below.

As early as 1970, Sandkull (Ref. 11) had identified distinctive properties of innovative organisations, i.e.

Open communication

Outside contact encouraged

Originality and eccentricity are encouraged

They are decentralised

Risk takers are encouraged

Diversity of communication is encouraged

In Hawthorne's book, "The Management of Technology" (Ref.12), he talks about "the need to develop management skills". He also discusses the management systems/operations factors, i.e. "Management systems tend to assume there will be no change and basic organisation theories ignore project work.

Dynamic people and organisations change in step with project characteristics of its technology, market and environment. There is a continual change in attitude."

"A firm's technological adaptability depends heavily on how a firm's organisational structure modifies itself ... to achieve the correct balance between 'hard and soft' information systems."

Gee, in his book "Managing Innovation" (Ref.13), lists the points to be considered by a company organising for innovation. These include:

1. Establish corporate objectives.
2. Ensure all members understand all objectives relevant to them.
3. Signal clearly management expectations.
4. Create a climate for innovation to achieve 3.
5. One man to be the focal point for responsibility.
6. Hold initial organisation to a minimum.
7. Allow organisational growth only as needs require.
8. Minimum interfaces between organisational levels and units.
9. Short communication lines.
10. Encourage informal discussions.
11. Reward success and understand failure.
12. Innovation means change - people, not structure, are the key.

These twelve points give a practical guide for creating the "climate for change".

Practical Implications

All the factors we have talked about throughout this paper, i.e. finance, technology, people, and management, all need to be included in the appraisal for the introduction of new technology, if discontinuities are to be avoided.

The financial and technological appraisals are comparatively straightforward to conduct. There are many 'standard' financial criteria in common use today, such as Payback period, Return on Capital, etc., and this appraisal is becoming more detailed with the advent of computer packages such as IVAN, which although it has drawbacks, is certainly a step in the right direction.

There are many reasons for a company wanting to introduce new technology - new products, new markets, increased flexibility, and so on. Once these reasons have been fully defined, it is a relatively simple process for the possible options to be investigated until the final choice is made.

The question of the appraisal of people and management is more difficult. There are many areas to be considered, such as the training, re-organisation of the workforce and management, integration of the technology into the existing operation, and so on. Different types of company will give more consideration to particular areas, depending on their size, products, etc.

The results from the questionnaire will show which factors will be more crucial to a certain type of company, and hence the important areas for a company to consider in their appraisal of the new technology. The results will also give some indication of who should conduct these appraisals.

We believe that if such a comprehensive appraisal as discussed throughout this paper were to be conducted for the introduction of new technology, this would result in the company having the greatest possible chances for success.

REFERENCES

1. C.E. Hutchin, "A Study of the Methods Employed in the Introduction of Text-Processing Systems to Organisations", Leicester Polytechnic, 1986.
2. J. Bessant and W. Haywood, "The Introduction of Flexible Manufacturing Systems as an example of Computer Integrated Manufacture", Brighton Polytechnic, 1985.
3. G. Rosegger, "The Economics of Production and Innovation", Pergammon Press, 1986.
4. R.N. Foster, Innovation, Macmillan, London 1986.
5. H. Marlow, "Managing Change: A Strategy for our Time", Institute of Personnel Managers, London, 1975.
6. S.T. Parkinson and G.T. Avlonitis, "Management Attitudes to FMS", Proc. 1st. Int. Conf. on Flexible Manufacturing Systems, 1982, IFS, pp. 405 - 412.
7. J. Adair, "Action Centred Leadership", MrGraw Hill, London 1973.
8. P. Primrose and R. Leonard, "Evaluating the Intangible Benefits of Flexible Manufacturing Systems by use of Discounted Cash Flow Algorithms within a Comprehensive Computer Program" Proc. Instn. Mech. Engrs., Part B, Vol.199, 1985.
9. A. Rusecki,"New Manufacturing Systems", University of Liverpool, 1987.
10. "Meeting the Challenge of Change" - Guidelines, Work Research Unit, A.C.A.S., 1982.
11. B. Sandkull, "Innovative Behaviour of Organisations", Lund, Sweden, 1970.
12. E.P. Hawthorne, "The Management of Technology", MrGraw Hill, London, 1978.
13. E.A. Gee,"Managing Innovation", Wiley, London, 1976.

Innovation: Adaptation and Growth, edited by R. Rothwell and J. Bessant
Elsevier Science Publishers B.V., Amsterdam, 1987 — Printed in The Netherlands

PROMOTING TECHNOLOGY TRANSFER: A PUBLIC POLICY PERSPECTIVE*

S.D. REID[1] and L.J. REID[2]

[1]Faculty of Administration, University of New Brunswick, P.O. Box 4400, E3B 5A3, Fredericton (New Brunswick, Canada)

[2]Vice-President, Reid Associates, 125 Charing Crescent, E3B 4R8, Fredericton (New Brunswick, Canada)

SUMMARY

This paper reports on a Canadian study conducted to identify strategies that policy makers should use to encourage license adoption among manufacturing firms. It provides an analysis of the factors implicated in license transfer and use and identifies major considerations that policy makers should take into account when promoting licensing as a means of technology transfer.

INTRODUCTION

Licensing is a low cost method for firms to acquire new and relevant proprietary knowledge from external sources. This form of technology transfer can be used by manufacturers to improve competitiveness, extend product lines and/or expand customer markets. Consequently, promoting license adoption among manufacturing firms represents an appropriate strategy for use by government policy makers to encourage industrial development.

Unfortunately, little information exists on the factors that influence license adoption by manufacturing firms. This paper addresses this deficiency by examining the licensing adoption behaviour among small and medium sized manufacturing firms. The investigation is confined to proprietary knowledge acquisition, in the form of inventions, process techniques, trademarks and patents. The paper identifies the scope and potential for licensing and elaborates on the role of licensing in firm performance.

LITERATURE OVERVIEW

A survey of the literature on licensing shows that:

- The import of advanced technologies through licensing facilitates development and modernization of manufacturing sectors.[1]
- Firms and industry sectors vary in their potential for exploiting proprietary knowledge acquired through licensing.[2]

*Acknowledgements to the Canadian Federal Department of Regional Industrial Expansion for sponsoring the study.

- Large firms often have technologies which they do not exploit directly, given the relatively limited commercial value involved. These technologies are often licensed to other firms.[3]
- Licensing, in the form of market or technical knowledge, is an alternative vehicle for encouraging new venture formation and firm expansion.[4]
- Manufacturers frequently use their suppliers as a prime licensing source.[5]
- Proprietary technology from university and government laboratories is acquired by the private sector through licensing to establish new manufacturing ventures.[6]

The above findings suggest that licensing has significant potential as a vehicle for promoting industrial development. However, policy makers require more specific knowledge of the license adoption process before they can usefully exploit it in this manner.

Consequently, information is required on: 1) the scope and extent of licensing activity among manufacturing firms, 2) the factors associated with licensing, 3) which firms represent the best targets for licensing initiatives, and 4) the mechanisms that encourage license adoption and use.

RESEARCH QUESTIONS

This paper addresses the following general questions:

1. What is the scope, extent and condition of current license use among manufacturing enterprises?
2. What factors favour licensing use by such enterprises?
3. Which specific industries in the manufacturing sector offer the greatest potential for systematic development of a licensing recruitment and technology transfer strategy?
4. What kinds of licenses should be targeted and license arrangements sought, by industry type?
5. How can public policies effectively and efficiently promote and influence licensing activity in the manufacturing sector?

RESEARCH DESIGN

This study was conducted during March-April, 1986. The firms under investigation represent the complete population of New Brunswick manufacturing firms listed with the Department of Commerce and Technology (Government of New Brunswick). Data was obtained through a self-administered bilingual (French and English) questionnaire sent by mail to 1,057 firms.

The questionnaire elicited information in three major categories. These are:

1. Firm Demographics: location, industry type, age, sales, economic performance and employee skill distribution.
2. Extent of Licensing: (a) knowledge, (b) search, and (c) interest in being

contacted about government licensing initiatives.

3. License Use: including license type, age, source, payment method and extent of contribution to gross annual sales.

RESEARCH FINDINGS

230 firms responded to the study, providing a 22% response rate. These firms were concentrated in the food and beverage (22.3%), wood (19.6%), miscellaneous manufacturing (18.3%) and metal fabrication (13.0%) industries, with the remaining respondents spread across 15 other industries.

Firm Performance Characteristics (see Exhibit I)

Responding firms are disproportionately larger than the population. 43.4% of responding firms have sales of $1 million (Cdn.) or more, in comparison to the population in which 74.5% have less than $1 million (Cdn.) in sales. Respondents are typically well-established; two-thirds are older than seven years of age and have more than $½ million (Cdn.) in annual sales. ¼ report that exporting accounts for more than 10% of total sales (see Exhibit I).

Firms with acquired licenses have distinctly higher annual sales: 65.5% report a gross turnover of $1 million (Cdn.), compared with only 40.7% of firms without such licenses with similar sales. Firms with an interest in being contacted about licensing tend to be generally much smaller in size. 44% of such firms have less than $½ million (Cdn.) in sales.

Most responding firms, irrespective of whether they have acquired licenses or not, report their export sales have not increased by 20% or more since 1983. However, there are major differences between firms with and without licenses in other performance areas. For example, firms with acquired licenses show a greater increase in employees, sales and new product introduction since 1983, than firms without licenses. Firms with licenses also show a greater propensity to internally develop patents,and intent to manufacture new products within the next two years.

Interestingly enough, firms that wish to be contacted about licensing initiatives have better performances overall than firms with no interest in being contacted. Firms in the former category are much closer in performance to firms with acquired licenses. This finding suggests that interest in license use is more likely to be found among firms that can be typically called dynamic.

Responding firms also sell other firms products (43.3%), either as distributors, agents or representatives (31.8%) or as resellers. However, these selling activities occur more frequently among firms with acquired licenses and/or an interest in being contacted about licensing initiatives than other firms. This finding indicates that a significant number of manufacturing firms also offer products they do not manufacture.

Nearly half of the firms (49.8%) report formally trained personnel in

Exhibit I: Comparison Among Responding Firms Using Selected Variables (N=230)

Firm Characteristics	ALL RESPONDING FIRMS: Without Acquired License (N=201)	With Acquired License (N=29)	With* License Interest (N=100)
1. Age:			
- Less than 4 years	13.5%	17.2%	18.0%
- Older than 7 years	67.7	62.1	66.0
2. Annual Sales:			
- Less than $½ million	39.3	17.2	44.0
- $1 million or more	40.7	65.5	36.0
3. 1985 export sales/ total sales greater than 10%	25.5	27.5	20.0
Firm Performance			
4. Increse in employees by 10% since 1983	44.8	51.7	48.0
5. Increase in sales by 20% or more since 1983	43.8	55.2	50.0
6. Increase in export sales by 20% or more since 1983	19.9	20.7	20.0
7. Introduced new products/ processes since 1983	44.8	55.2	51.0
8. Plan to manufacture new products within next 2 years	54.7	75.9	66.0
9. Currently have patents etc. that were developed internally	31.3	41.4	35.0
Type of Operation			
10. Sell other firms' products	43.3	48.3	52.0
11. Act as distributor and/or manufacturer's agent for other firms' products	31.8	44.8	37.0

*Also Includes Firms With Licenses

management, accounting or marketing functions. However, firms report little presence of such personnel in other areas more directly related to the manufacturing function. For example, a minority report the presence of formally trained personnel in manufacturing or production (29.9%), process and quality control (17.9%), diemaking and machinists (11.4%), computer aided design and/or manufacturing (6.0%), instrumentation (.5%) and microelectronics (.5%).

In general, the findings on firm demographics and performance show that licensing activity is associated with: 1) increased employment and sales, and 2) a higher propensity to: (a) introduce new products, (b) internally develop products, and (c) expand manufacturing capability. A significant minority of these firms also sell products that they do not manufacture. In addition, licensing activity is more prevalent among larger well-established firms with a higher level of formally trained employees in varied production areas.

Extent of Licensing Knowledge, Search and Interest

The majority of responding firms without an acquired license (70.2%) indicate they have very little or no knowledge of the ways in which licensing can benefit their firms. Surprisingly, many firms (62.1%) involved in licensing also express similar views. Firms interested in being contacted about licensing, when compared with firms without licenses, show little differences in their knowledge about licensing, but conduct marginally more license searches.

Firms with licenses show higher incidences of license searches during the past three years. 49% of these firms had actively searched for licenses within this period, in contrast to 25% of firms without a license. Passive searches for licenses include coming across potential licenses occasionally or engaging in non-systematic license searches. In spite of these generally limited searches, nearly half of all responding firms express an interest in licensing information and in being contacted regarding possible licensing initiatives.

Firms primarily search for licenses which can provide products that they currently manufacture and, to some extent, for manufacturing products that they now sell or that are entirely new to their product lines. These findings suggest that the primary purpose for acquiring a license is to secure proprietary rights applicable to products/markets with which firms are familiar, in contrast to engaging in totally new products/markets.

Trade shows, personal industry contacts on a national and international basis, government agencies and industry publications, in descending order, are the sources most frequently used by firms to secure licensing information. These sources can be categorized according to the specific information they convey about particular licenses and licensing topics. Their selective use by firms suggests that firms favour licensing information sources that are both direct and specific.

Licensing Strategies and Use

Most firms (62.1%) with acquired licenses report receiving licenses independently from third parties. This suggests that the remaining firms (37.9%) have acquired licenses in non-arms-length transactions, probably through subsidiary agreements. 19 of the 29 acquired licenses were obtained by licensees in their roles as distributors, manufacturers' agents and/or representatives.

This finding is consistent with previous evidence showing that firms with licenses have a greater propensity to sell other firms' products and act as distributors/manufacturers' representatives, than firms without acquired licenses. These results indicate that manufacturing, through licenses acquired from suppliers, is the most frequently used licensing approach. This method of license acquisition conceivably continues as manufacturers pursue expansion into new product areas.

Exhibit II: Comparison of Extent of Licensing Knowledge, Search and Methods Across Responding Firms (N=230)

	ALL RESPONDING FIRMS		
Licensing Knowledge & Search	Without Acquired License (N=201)	With Acquired License (N=29)	With* License Interest (N=100)
1. Very Little or No Knowlege	70.2%	62.1%	69.0%
2. Little or No License Search	75.1	51.7	68.0
3. Seek Licenses for Products:			
a) Currently Manufactured	14.9	34.5	20.0
b) Currently Sold	4.5	13.8	10.0
c) Niether Manufactured Nor Sold	6.5	10.3	8.0
Licensing Information Sources Used			
1. Trade Shows and Exhibitions	9.0	34.5	13.0
2. Canadian Federal/Provincial Government Sources	10.0	24.1	13.0
3. Personal Contact with Firms in Same Industry:			
a) Canadian	8.5	24.1	8.0
b) U.S.	5.5	13.8	8.0
c) Other Foreign	3.0	10.3	6.0
4. Industry (Trade) Publications	5.5	13.8	9.0
Extent of Licensing Interest			
1. Interested in Licensing Info.	42.8	44.8	86.0
2. Interested in Being Contacted	45.8	51.7	100.0
License Source & Use			
1. Acquired from Other Firm or Agency	n/a	62.1	8.0
2. Acquired through Distributor/ Agent/Representative Route	n/a	65.5	11.0
3. Used to Manufacture New Products	n/a	41.4	11.0

*Also Includes Firms With Licenses

Firms show considerable variation in the: (a) type of licenses they have, (b) sources from which these are acquired, and (c) type of payment method used. 29 firms had acquired 41 licenses, with product and process manufacturing licenses accounting for 61%, patent and trademark 17%, brand names 7%, and technical and engineering services 5%. This aggregate distribution is somewhat misleading, since 10 firms account for nearly half of the acquired licenses, including one firm which has five manufacturing franchise agreements.

Product and process manufacturing licenses and patents form the largest category of licenses acquired during the last three years. In contrast, older licensing arrangements typically consist of trademarks, brand names, technical and engineering services. These findings suggest that there is an evolutionary pattern of licensing use, which begins by securing simple licenses, then acquiring more advanced and/or sophisticated license arrangements as time passes.

Canadian, U.S.A., and other foreign (European) companies are the main licensors, comprising 44%, 27% and 12%, respectively, of all reported licensing sources used. The National Research Council is the only other license source used by respondents. No responding firms report university and/or public research laboratories as licensing sources.

Payment Methods

Little consistency exists in licensing payment methods across responding licensees. Royalties, one-time payments, annual fixed fees and percentage of sales are the most frequently used single methods for license payment. The data provided on 36 licenses show that firms use royalties (10/36), annual fixed fees (4/36), one-time fixed fees (4/36), percentage of sales (3/36), combinations of these preceding alternatives (8/36), and special payment methods involving either cross-licensing and/or marketing arrangements (2/36). Firms with more than one license generally use different payment methods for each license. These payment characteristics are also generally true for the subgroup of licensees desiring further contact about licensing.

Only 19 firms provide information on the turnover derived from licensing. Eight firms indicate that products manufactured under license account for more than ¼ of their gross annual sales, while four firms indicate that such products contribute between 11%-25% of gross sales. The remaining seven respondents derive 10% or less of their gross annual sales revenues from goods manufactured under license. A similar pattern of sales distribution is also found among the subgroup of licensees who indicate their interest in being contacted.

IMPLICATIONS

Licensing activity and interest is generally dispersed in proportion to the manufacturing firm population's industry distribution. Industrial sectors containing metal fabrication, miscellaneous manufacturing, food and beverage and wood account for nearly half of all acquired licenses and three-fifths of all firms interested in being contacted about licensing initiatives. Consequently, these sectors should represent initial targets for licensing initiatives. The requirements of these and other manufacturing groups should be recognized in sector-specific strategies which reflect both the differences in industry populations and their production characteristics.

The higher economic performance shown by firms with acquired licenses is significant. These firms, as a group, show greater increases in employees and sales, product introductions and internally developed patents. They are also likely to be more diversified and have higher levels of formally trained employees than manufacturing firms as a whole. Although it is not clear whether licensing propensity is the cause of these performances or the reverse, it is clear that dynamic enterprises are over-represented among firms having acquired licenses and/or those interested in being contacted about licensing initiatives.

A substantial number of responding firms act as distributor representatives, agents and/or resellers of other firms' products. Acting in this capacity, they are in a position to develop and assess new market opportunities and segments which may eventually prove viable for manufacturing on own-account. This type of firm behaviour provides key clues to the license introduction and adoption process.

The research findings show low levels of licensing activity among manufacturing firms, but indicate a relatively high interest in being contacted about licensing initiatives. The limited license information search and expressed ignorance of licensing benefits indicate a need for strategies aimed at educating the manufacturing sector about what licensing is, what the process involves and the potential benefits it can confer.

It is important to note that personal contacts, trade fairs and exhibitions and government agents/agencies, are the main licensing information sources used. Consequently, strategies aimed at increasing license adoption should assign a primary and explicit role to these vehicles in disseminating licensing information.

Many of the firms expressing a desire to be contacted regarding licensing initiatives are small, with sales under $½ million. The resource profiles of these firms, as a group, do not match those of firms already engaged in licensing. As a result, these firms cannot be said to represent prime candidates for license acquisition. However, smaller firms have alternative routes for expansion through acquisition of distributor, agent and/or manufacturing representative rights for allied products. This initial step in the licensing process must be recognized and emphasized in public policy attempts to stimulate technology transfer.

Responding firms with acquired licenses, have no consistent method of license arrangement, sourcing and/or payment. These results underline the use of licensing as opportunistic, rather than a systematic process of strategic development and planning. Considerable emphasis needs to be placed on educating firms about the variety of licensing arrangements that exist, the various payment methods available and the situational flexibility such arrangements often permit. This focus is likely to help firms whose limited licensing knowledge may deter them from going the licensing route.

A substantial minority of firms show the presence of internal proprietary knowledge, in the form of developed patents or products. This knowledge represents a resource that can either be traded on own-account -- outward licensing -- or be used to secure other licenses through cross-licensing agreements. The potential for such activity seems considerably high and offers evidence of an alternative type of licensing activity, not normally the concern of public policy.

LIMITATIONS OF STUDY

Several factors should be taken into account in evaluating this study's validity and interpreting its results. These factors relate to: 1) population census accuracy, 2) differential response rates across industry sectors, and 3) differences between respondents and non-respondents.

Total responses include seven firms which had ceased operation or relocated and eight firms that indicated they were incorrectly classified as manufacturers. It is difficult to ascertain the extent of misclassification in the population, since this can only be indicated by survey respondents.

Differential response rates were observed across geographic locations and industries, with firms in urban areas accounting for 31.7% of all responses, although representing 56.4% of the manufacturing firm population. In addition, firms in food and beverage, wood, metal fabrication and miscellaneous manufacturing sectors account for 73.2% of all responses, while they represent 63.4% of the firm population.

Responding firms are also disproportionately larger than the population. 43.4% of responding firms have sales of $1 million or more, in comparison to the population which has 74.5% of all manufacturing firms with less than $1 million. This difference between respondents and non-respondents limits the statistical generalizability of the study. However, it also argues for its pragmatic validity, since these respondents are a valid group for the specific purpose of identifying the extent of and potential for licensing.

CONCLUSION

This study shows that encouraging license use should be considered by government policy makers as a viable strategy for stimilating industrial development. It indicates that policy initiatives aimed at promoting license adoption among manufacturers should:

1. Create awareness of licensing and its benefits among manufacturing firms.
2. Establish support services that specifically assist in the search for appropriate license and product market expansion alternatives.
3. Provide facilities and services that envourage optimum license adoption and use.
4. Target industry sectors, rather than use a general industry approach.

REFERENCES

1 Ozawa, T., Transfer of Technolgy from Japan to Developing Countries, Report #7, New York, N.Y.: United Nations Institute for Training and Research,1971.

2 Rothwell, R. and W. Zegfeld, *Industrial Innovation and Public Policy*, Westport, Connecticut: Greenwood Press, 1981.
The State of Small Business: A Report to the President, U.S. Small Business Administration, Washington, D.C.: U.S. Government Printing Office, 1983.

3 Crookell, H., *Getting Along Without Multinational Firms*, Paper presented at the Academy of International Business Conference: Honolulu, Hawaii, December, 1982, pp. 18-20.

4 Beckman, C.C., *Developing New Export Markets*, Ottawa, Ontario: The Conference Board of Canada, 1981.

5 Wills, R.M., *The International Transfer and Licensing of Technology in Canada*, Technological Innovation Studies Research Report No. 83, Ottawa, Ontario: Department of Regional Industrial Expansion, 1982.

6 Cameron, Joyce, "UNB Prof's Mapping System Fills Many Bills," *Innovation*, September/October, 1985, pp. 8-9.
"Westmorland Lab Has New Idea," *The Brunswick Business Journal*, January 4, 1985, p. 16.

Innovation: Adaptation and Growth, edited by R. Rothwell and J. Bessant
Elsevier Science Publishers B.V., Amsterdam, 1987 — Printed in The Netherlands

IMPROVING THE CHANCES OF NEW PRODUCT SUCCESS

I.BARCLAY and M.H.BENSON

Department of Industrial Studies, University of Liverpool, P.O.Box 147, Liverpool, L69 3BX.

(This research is jointly sponsored by the SERC. and Delco Electronics Overseas Corporation, Moorgate Road, Kirkby, Liverpool, L33 7XL.)

SUMMARY

In the past thirty years a large number of research studies have probed into the new product development process and identified the attributes associated with new product success. Despite this research however few companies are aware of the results. New product failure rates remain high and the problem has become not one of identifying what makes a new product successful, but one of applying the lessons that have already been learnt.

One of the reasons for the lack of progress is probably due to the way in which the results have been presented. The methods for applying the research offer mainly "one off" solutions that will only add more rules and procedures to the formal structure of a company. Also, few companies have a system that will allow them to adapt the management of new product programmes to changing demands and needs.

This paper proposes that Organisation Development is one of the most effective ways for companies to overcome these problems. It will allow them to make use of the lessons learnt from new product success, and react to changing demands. Organisation Development is a continuous process and the first stages involve data collection. A method is presented that will allow companies to collect this data and compare their own new product organisation with the attributes that have been identified as being important to success.

INTRODUCTION

New product development is perhaps the most important, yet least understood activity within a company today. It is crucial to the long term survival of manufacturing industry and remains one of the most important routes to corporate growth. New product development is a complex process with inputs from a large number of disciplines. It requires the coordination and collaboration of many people across various functional boundaries. As many of the disciplines managed today become steady state operations, new product development remains a creative and fast moving activity. New products developed have to adapt to new markets, more aggressive competition, new technologies and new management abilities.

If the new product development process is to be managed effectively and efficiently, the techniques used and emphasis put on them must change in accord with the changing new product demands. Companies need to look closely at how they develop new products, and take time to improve the management of this process.

The paper is in three parts; the first reviews some of the new product research that has tried to identify the reasons behind the success and failure of new products. The second questions the lack of awareness and application of this research amongst managers of new product development. Finally a method is proposed that will allow companies to effectively apply the research, and hopefully build on the lessons that have been learnt.

RESEARCH INTO NEW PRODUCT SUCCESS AND FAILURE

Over the past 30 years a significant amount of research has been carried out to try and identify the reasons behind new product success and failure. The initial work, carried out during the late 50's and 60's, drew conclusions from individual new product case studies. As the research progressed larger surveys were carried out; these were, in effect, groups of individual case studies. In 1957 Carter and Williams (ref. 1) identified twenty four characteristics of technically progressive firms, this was based on a sample of 152 industrial case studies. Marquis (ref. 2), in 1969, analysed over 500 innovations in 121 companies before drawing together the characteristics of successful innovations. Similarly Langrish (ref. 3), in 1972 published a list of factors associated with a firms success in technological innovation.
This research, and other early work, concentrated on identifying the reasons associated with the success of innovations and especially new products.

It was not until the early 70's that research began to compare success with failure. One of the first studies to do this was project SAPPHO (ref. 4). Project SAPPHO compared pairs of innovations from the same market. This study concentrated on the instrumentation and chemical industry. A "pairwise" comparison technique was used to differentiate between successful and unsuccessful policies and practices associated with successful and unsuccessful innovations. Results from the second phase, based on a sample of 43 paired comparisons, identified five main areas of difference between success and failure. Successful innovators were found to:-

1. Have a much better understanding of user needs.
2. Pay more attention marketing.
3. Perform development work more efficiently.
4. Make more use of outside technology and advice.
5. Have key, responsible individuals, in greater positions of authority.

The techniques used by SAPPHO for comparing success with failure were a major breakthrough and have become the choice of many researchers since.

One of the largest and most comprehensive studies to date was Project NewProd. (refs. 5-7) carried out by Cooper in the late 70's. A mailed survey was used on a sample of 195 industrial manufacturers active in new product development. Cooper identified three "keys" that were strongly associated with new product success:-

1. Having a unique and superior product in the eyes of the customer.
2. Having a strong market knowledge and proficiency.
3. Having technical and production synergy and proficiency.

These results differed from SAPPHO in that they were oriented more towards product and market characteristics associated with success and failure. SAPPHO on the other hand put more emphasis on the managerial and organisational characteristics associated with the new product development process.

These differences in the results are probably due to differences in methodologies used. In 1982 the Stanford Innovation project (ref. 8) was carried out by Maidique in order to overcome these differences. Maidique identified eight broad areas that appear to be important to new product success in a high technology environment.

1. Market knowledge gained through frequent and intense customer interaction.
2. Planning of the new product process, especially the R&D phase.
3. Coordination of the new product process, especially the R&D phase.
4. Emphasis on marketing and sales.
5. Management support for the product throughout the development and launch stages.
6. The contribution margin of the product.
7. Early market entry.
8. Proximity of the new product technologies and markets to existing strengths of the developing unit.

A more recent study carried out by Souder (ref. 9) identified the factors that were found to influence the success and failure outcomes of new product innovations. Souder noted that a high degree of understanding of the technical problems and the users' needs, a high degree of fit between the technology and the company's level of expertise, and a high quality of resources, were repeatedly found to characterise the most successful projects.

The above outline of some of the major studies that have been carried out will give just a flavour of the type of results obtained. There are many other studies, not mentioned here, that have identified similar reasons behind new product success. A common agreement amongst researchers in this area is that there is no one panacea, or solution, that will improve new product success. The factors that affect new product success and failure are complex and will vary between different industries, organisations, products and so on. Despite this however the results are strikingly similar, especially when consideration is given to the complexities associated with measuring new product success. For example, studies mentioned in this paper have used different samples and sample sizes, different methodologies and even different definitions of success.

AWARENESS AND APPLICATION OF NEW PRODUCT RESEARCH

The lack of awareness of the research.

From an initial pilot study of the approaches used by new product managers, it was found that few managers had actually come across this work and were aware of the findings. It is our contention that the results discussed above are not widely known by the people involved in new product development. To test this hypothesis in more depth we included in a recent questionnaire a section that referred specifically to the respondents knowledge of the research studies (The scope of the questionnaire as a whole was much greater than results reported in this paper.) The questionnaire was sent to various engineering companies in the UK., and was aimed at the director, or manager responsible for new product development. The results reported here are based on our initial findings.

Managers were first asked to indicate if they had heard of any of the following studies into the success and failure of new products:-

	% managers who replied YES
(1) Project New Prod.	0%
(2) Project SAPPHO.	2%
(3) Stanford Innovation Project	6%
(4) Booz Allen & Hamilton	16%
(5) Langrish (Wealth from Knowledge)	0%
(6) Others (specified)	2%

Managers were also asked if they had applied any of the studies; only 10% of the managers who had heard of the research had taken steps to apply the results. We are thus left with two conclusions:-

1. The researched evidence on the success and failure of new

products is not widely known.

2. Those managers who are aware of the research, in general, are not applying it to the management of new product development programmes.

Application of research into new product success and failure.

From the research into new product success and failure, no clear methodologies have been put forward on how to apply the results and reap the benefits of the findings. The majority of the results have been presented as lists of variables that affect new product success. Cooper (ref. 10) highlighted this lack of application as a major problem in 1983 and proposed a flow model approach to the development of new products. Cooper went on to suggest the redesign of a company's new product development process into the form of a model that took into account the research findings. In an earlier paper Cooper (ref. 11) suggests that an effective strategic plan is essential. Rubenstein (ref. 12) suggests changes to the new product policy. Souder, (ref. 6) based on a large database of 289 new product development innovations, presents ten guidelines for managers involved in new product development. The majority of the studies however, including SAPPHO and the Stanford Innovation project, present the results simply as lists of key findings.

Of the methods that have been put forward as ways of applying the research all will improve the management of new product programmes, but there are some major problems with these types of solution:-

1. The methods presented are in effect "one off" solutions, that is they do not take into account any changes that will affect the new product development process once the solution has been implemented.
2. They focus on only one aspect of the new product development process, and do not give a balanced view.
3. Companies may not have an organisational structure or culture that will allow improvements and changes to be implemented.

The research presented in this paper aims to develop and test methods that will improve the chances of new product success. Current improvements will tend to be reactions to problems rather than anticipation of changes to improve the process before it goes wrong. In todays rapidly changing environment companies need to be able to apply research that has already been carried out. Effective application of this research will mean substantial changes to the organisation. These changes may alter the development processes, management abilities, technical capabilities etc. It is essential therefore that these changes can be properly implemented.

The Management of Change.

A large amount of research has been carried out into the management of change. In 1970 Sandkull (ref. 13) published work on the distinctive properties of innovative organisations. He concluded that innovative behaviour is a manifestation of morphogenic process (change) whereas project realisation is morphostatic (steady). These give rise to conflict and it is therefore imperative that people are included in the description of an innovative organisation. In 1975 Marlowe (ref. 14) presented a strategy for managing change.

Much of this and other similar work was being carried at a time when technology was relatively stable. Designers had time to engineer products to a relatively stable manufacturing base. Today however technology both within the product and the process is changing rapidly. The advent of new manufacturing systems will often mean that the designer is facing a changing manufacturing base. The problems of managing this constantly changing situation have been summarised by Hawthorne (ref. 15) "Management systems tend to assume there will be no change, and basic organisational theories ignore project work. A firm's technological adaptability depends heavily on how a firm's organisational structure modifies itself."

Since new product development programmes are very often at the forefront of change, and application of any research factors will undoubtedly involve change it is essential that any method of improving new product programmes takes into account factors influencing the success of these changes. Managers need not only to create an atmosphere conducive to innovation but one which can also adapt and change as and when required.

In order to improve the management of new product programmes and the chances of new product success, improvements must be:-

1. Made with an awareness of the factors that lead to new product success and failure, so that these factors can be created or improved within the company.
2. Continuous, allowing the new product programmes to react to changes in the environment.
3. Combined with other methods of improving new product programmes, such as improving the strategy, so as to provide a balanced approach to new product development.

This research puts the view that if new product failure rates are to be reduced and the chances of new product success improved, the most effective way is to create the above conditions. These conditions can be created using the technique of Organisation Development.

FOCUSING O.D. ON NEW PRODUCT DEVELOPMENT

Organisation Development

Organisation Development (O.D.) as defined by Luthans (ref. 16) is a "systematic way of managing change". Bennis (ref. 17) gives a more detailed description, and refers to O.D. as "a response to change, a complex educational strategy intended to change the beliefs, attitudes, values and structure of organisations so that the can adapt to new technologies, markets and challenges and the dizzying rate of change itself."

Woolhouse (ref. 18) says there is a need for Organisation Development in many situations, and of four situations put forward as needing O.D. the following three are directly relevant to new product development:-

1. Allowing existing organisations to adapt more easily to changes in the environment.
2. Where the organisation is failing to accomplish its objectives in terms of output, quality or profitability and where the nature of the organisation itself is limiting the performance of those who work in it.
3. Where the adoption of new technology, systems or methods requires corresponding changes in the organisation for proper implementation.

From the above it can be seen that O.D. is an ideal method for improving the management of new product development because:-

1. New product programmes must react to changes in the environment such as technology advances, changing customer needs and more aggressive competition.
2. The organisation may be failing to achieve the level of new product success and the new product management and organisation itself may be limiting the chances of new product success.
3. The adoption of new product technology and the development of systems and methods are essential to efficient new product development.

The Organisation Development process.

In its basic form O.D. is regarded by most practitioners as a four stage process:-

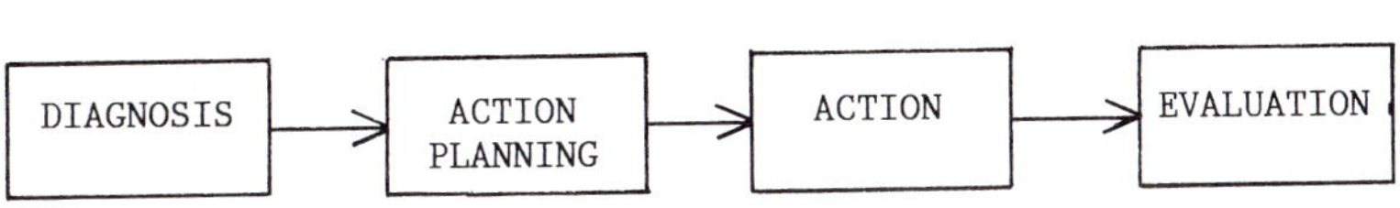

Every company is unique and must therefore shape the above process to fit in with its own culture and structure.

Using O.D. to improve new product development.

To improve the new product development process O.D. needs to be implemented through a multi-disciplinary team so that experience can be drawn from the various stages of the new product development process. One of the first priorities will involve the development of the team itself. This is best achieved through an "off site" team building programme. The objective of the programme being to build the team into a cohesive unit with a basic understanding of O.D. principles, and the team's role in improving the management new product programmes.

Diagnosing the "real" problems.

Once the team is established it is critical to diagnose the real problems facing new product development. Baker and Gorman (ref. 19) for example, conclude that diagnosis is the key to O.D. effectiveness. It is at this stage that the team needs to draw on the research into new product success and failure. It is also essential that the team identify the real problems facing new product development and not those as perceived by an individual department, executive or manager. In order to do this effectively the team must collect information from several sources. This should include:-

1. The factors associated with the success and failure of new products.
2. Key executives perceptions of the problems.
3. The teams own particular experiences.
4. The views of other managers and staff involved in new product development programmes.

This information can be obtained through two methods:-

1. A new product survey
2. Interviews

The new product survey.

A team has been established in the company sponsoring this research so as to test and evaluate the use of O.D. to improve new product development. The survey was developed to aid the team in diagnosing the real problems facing new product development. It is based on the findings of the major research studies into new product success and failure, as reviewed in the first section of this paper. The factors found to be associated with new product success were listed, common ones merged into groups, and then broken into 5 main areas. These are:- (in no order of priority)

1. A unique and superior product that clearly meets customer wants and needs.
2. Good communications with efficient planning and collaboration between the individuals and departments involved.
3. A good synergy between the new product being developed and the company in the area of technology and production, and a strong proficiency in undertaking technical and production activities.
4. An open minded and professional management who can create an atmosphere of trust, coordination and control.
5. A strong market knowledge and strategy combined with a clear understanding of user wants and needs.

Each of these areas represents a section in the survey. In each section there are around eight qualities that are associated with new product success. The idea of the survey is that members of the organisation respond by either agreeing or disagreeing (on a scale of 1 to 5) that a particular quality is present inside the organisation.

In this way the survey has the double advantage that it makes people aware of the research findings (education) and, at the same time, provides the team with a sound database on which to base the action planning stage. Key problem areas can be followed up by more in depth interviews, carried out by the team members. The factors relevant can then be analysed and qualities needed identified and built upon.

The results of the survey should be fed back to members within the organisation so that they become aware of the research that is relevant and important to the company.

The Action Planning and Action stages.

With a thorough diagnosis completed the team can then decide on the appropriate form of action. This action planning stage needs to be carefully thought out so that if possible it involves the people who will be affected by the changes. During the ACTION stage, changes will be made and tested, with modifications to ensure the success of the programme. The evaluation stage should review the programme against the criteria agreed at the outset. A particular programme may end but the process itself will not. Organisation Development should build on experience, continually allowing the management of new product programmes, and the organisation, to adapt to the changing environment, technology and people needs.

REFERENCES

1. C.F.Carter & B.R.Williams, Industry and Technical Progress; factors governing the speed of application of science. Oxford University press, London, 1957.
2. D.G.Marquis, The Anatomy of Successful Innovations, Innovation, 1 (Nov. 1969) 28-37.
3. Langrish et al., Wealth from Knowledge, Macmillan, London, 1972.
4. R.Rothwell et al., SAPPHO updated - project SAPPHO phase II, Research Policy, 3, (1974) 258-291.
5. R.G.Cooper, Identifying Industrial New Product Success: Project New Prod., Industrial Marketing Management, 8, (1979) 124-135.
6. R.G.Cooper, The Dimensions of industrial new product success and failure, Journal of Marketing, 43, (Summer 1979), 93-103.
7. R.G.Cooper, Project New Prod: Factors in New Product Success, European Journal of Marketing, 14, 5/6, (1980) 277-292.
8. M.A.Maidique & B.J.Zirger, A Study of Success and Failure in Product Innovation: The Case of the U.S. Electronics Industry, IEEE Transactions on Engineering Management, EM-31, 4, (Nov 1984) 192-203.
9. W.E.Souder, Managing New Product Innovations, Lexington Books, 1987.
10. R.G.Cooper, A Process Model for Industrial New Product Development, EM-30, 1, (Feb 1983) 2-11.
11. R.G.Cooper, Strategic Planning for Successful Technological Innovation, Business Quarterly, 45, (1978) 46-54.
12. A.H.Rubenstein et al., Factors Influencing Innovation Success at the Project Level, Research Management, (May 1976) 15-20.
13. B.Sandkull, Innovative Behaviour of Organisations, Student Literature;Lund, Sweden (1970).
14. H.Marlowe, Managing Change: A strategy for our time, Institute of Personnel Managers, London, 1975.
15. E.P.Hawthorne, The management of technology, McGraw-Hill, 1978.
16. F.Luthans, Organizational Behaviour, McGraw-Hill, 3rd.Ed. 1981.
17. W.G.Bennis, Organisation Development its nature, origins, and prospects, Addison-Wesley, 1969.
18. J.Woolhouse, The Gower Handbook of Management, Gower, 1983.
19. H.K.Baker & R.H.Gorman, Diagnosis: Key to O.D. effectiveness, Personnel Journal, (Oct 1976).

Innovation: Adaptation and Growth, edited by R. Rothwell and J. Bessant
Elsevier Science Publishers B.V., Amsterdam, 1987 — Printed in The Netherlands

MANAGEMENT DEVELOPMENT AND TECHNOLOGICAL INNOVATION

A. Jefferson

SUMMARY

This paper develops the theme that Management Development, in order to assist in Technological Innovation, must be put in the context of the organisation in which it is undertaken.

The paper then examines the importance of the organisation's environment, the "vision" of the organisation and the culture of the organisation and demonstrates that it is necessary to fully understand these areas before considering Management Development.

Some key facets of the implementation of Innovations and the management of change are then examined.

INTRODUCTION

What I have attempted to do in this paper is outline an approach to relating Management Development and Innovation. I have tried to give a framework which should be useful for most organisations in attempting to develop a Management Development approach which is both appropriate for them and supports Innovation.

The starting point of any activity is an understanding of exactly what is being attempted. In this instance, the key starting point is to try to define Innovation. The definition chosen is one with which no doubt a number of you are familiar :

> "Industrial Innovation includes the technical design manufacturing management and communication activities involved in the marketing of a new (or improved) product or the first commercial use (or improved) process or equipment" (Freeman 1974) "(ref.1)".

Innovation, therefore, involves a long chain of events that link R&D departments at one end of the chain with the market place at the other end of the chain and should permeate every aspect of an organisation. Innovation is about change, often major change, not something that many organisations or many people find comfortable.

At this point I had better explain the process of Management Development. Management Development should be about providing the organisation with the managers it needs to succeed and compete both now and, equally importantly, in the future. This makes the heroic assumption that organisations know the answer to these questions and often they don't. Without these starting points, then attempting to develop a relevant and supportive Management Development strategy is difficult.

Management Development is best seen as an investment rather than a cost and it is worth remembering that any manager represents a very significant investment and that this investment needs maintaining and improving in much the same way as any other investment. Indeed, in many organisations, people are often the most neglected investment. Management Development is an important tool to enable both the organisation and individual managers to adapt to, and cope with, change.

I must also point out that Management Development does not consist solely of management training courses which forms only one part, and often a small part, of the total effort. Furthermore, Management Development is part of a much wider Human Resource area which has important implications for Management Development and includes:

(a) Recruitment

Does the organisation recruit the people it needs to succeed both now and in the future?

(b) Performance Appraisal

Does the organisation know the strengths and weaknesses, skills and abilities of its current management population? All too often, the answer is that it doesn't.

(c) Promotion and Succession

Amongst the management population, are there the people with the skills, abilities and motivation to fill more senior jobs in the future and how best can they be developed to be ready for their more senior roles? This may involve secondments or job rotation to give the required breadth of experience.

(d) Organisation Development

Organisation Development is best described as a long term effort to improve organisations problem solving and renewal process, particularly through a more effective integration of the organisation's management styles, structures, culture and business strategy. Obviously, this represents major long term effort and most organisations have not yet achieved this level of sophistication.

Given the breadth importance and pervasiveness of these areas, it is obviously impossible to outline the Management Development policy or approach that is appropriate to every organisation and situation. There is, however, an approach which I think can be adopted to enable each organisation to develop a Management Development philosophy to fit its own circumstances. The steps one needs to go through to do this are as follows:

(i) Relate the organisation to its environment.
(ii) Understand the organisation's vision, policies and strategies.
(iii) Understand the organisation culture.
(iv) Recognise the problems of implementing Innovation.
(v) Management of Change/The skills of successful Implementation of Change.

1. RELATE THE ORGANISATION TO ITS ENVIRONMENT

One of the most important factors in preparing any organisation for Innovation is to ensure there is a wide awareness of the competitive environment within which the organisation operates. Obviously, these environments can vary greatly from those which are undergoing almost continuous or sometimes discontinuous change to environments where the pace of change is much more gradual. This has significant implications for the nature of the organisation, the type of people in it and the style of management which is necessary to compete effectively. Where the pace of change is slow, then a slow moving bureaucracy and a very hierarchical management style can survive if not necessarily prosper. On the other hand, in a fast changing environment, an organisation which adopted that approach would be most unlikely to last very long. The best approach to this relationship between the Organisation and its Environment and the implications for strategy and management is in 'Implanting Strategic Management' by H. Igor Ansoff "(ref.2)".

What this highlights is that many of the key decisions about a Management Development approach go back several stages to the organisation structure and the type of people who work in that organisation.

One of the greatest challenges to any management is when the environment changes from being relatively stable to being relatively dynamic and the organisation has to adapt to fit the new circumstances. This will involve a <u>total</u> reorientation of the management or rather a total reorientation of those managers that survive.

Obviously this is a situation where a major Management Development effort is needed to ensure that both the organisation and the people in it can cope with the changes.

2. ORGANISATION VISION

Although this is currently very much of a buzz word or buzz phrase, nonetheless, it is of great importance and is the second building block in the approach to determining an appropriate Management Development approach for an organisation. Management Development is, incidentally, a powerful aid of either reinforcing or changing the organisation vision.

One of the keys, however, to any progress in this direction, is that the top management is clear on the vision and that the vision is appropriate for the environment in which the organisation operates. If the vision is not appropriate, then obviously there are major problems ahead for that organisation.

Stemming from this vision and consistent with it, would be the strategy the organisation adopts in order to compete in its market place or market places. The key role for management and <u>not</u> Management Development in this area is to ensure that the organisation's strategies or strategy and its implications are <u>fully</u> understood. In my experience, it is all too often the case that the understanding of the organisation's strategy and its implications is far less than the top management of the organisation believe and unless considerable efforts are made to communicate the organisation's strategy, then certainly at middle management levels, it will be either not understood or misunderstood. This means that parts of the organisation could be pulling in different directions.

A good clear vision can be a good motivator for management throughout the organisation and can give a clear sense of purpose and direction which everyone can relate to.

The organisation's vision and strategy is <u>the</u> key starting point both for Innovation and for Management Development. If the organisation at the highest level does not have a clear sense of purpose, as either an innovative organisation or an organisation where Innovation and change are important, then this will be reflected throughout the organisation. Without a clear sense of purpose or direction from the top, then Innovation will be far more difficult. No matter how good or how innovative the people in the organisation, swimming against the tide is always more difficult.

If the organisation has no clear strategy for Innovation and no clear place for Innovation in its vision, then it is difficult to see how Innovation can be successful on a large scale. It is even more difficult for innovative activity to be approached coherently and consistently throughout the organisation.

In much the same way, the organisation's vision and its strategy is the key starting point for its Management Development approach. It is also, therefore, the point at which the key aspects of the relationship between Management Development and Innovation are set. Management Development, like Innovation, does not, or should not, exist in splendid isolation as a "bolt-on" part of the organisation. The whole thrust of Management Development should be keyed in to what the organisation is attempting to achieve. Far too often, Management Development is not seen as an integral part of the organisation vision and strategy. This is even more important today when the quality of management is one of the few sustainable competitive advantages any organisation can have and innovation requires a higher calibre of management than stability.

What this means for Management Development is that the approach to be taken in all aspects from Recruitment through Appraisal to the actual training, has to be geared to the vision and strategy of the organisation. For each organisation, the approach adopted will tend to be different - in the same way as every organisation is unique in at least some ways, then Management Development must provide a unique approach to the problems and opportunities which confront the organisation.

What I am trying to argue against here is the "quick-fix" solution on the borrowing of an approach used elsewhere and applied in a different context. Not surprisingly "quick-fixes" and wholesale borrowing as an approach are no substitute for a more thought through and tailored approach.

3. ORGANISATION CULTURE

In organisation terms, culture can best be thought of as:

"the shared values and norms that permeate the organisation. These are largely based on past successes and are enshrined in the terms and strategies of the organisation and serve to reinforce continuity."

One of the biggest single problems that most organisations face is that they are inevitably dominated by the past and fail to recognise that the factors which led to success in the past may not be as appropriate for the future. This is especially true when the environment in which an organisation operates changes dramatically (which often seems to happen). One of the most important roles Management Development can play in an organisation is to ensure that the values and norms are continually questioned and exposed to analysis. Once again, this goes far wider than merely training courses but has significant implications for recruitment and promotion. Far too often, organisations recruit and promote clones of current managers. A better way of inhibiting Innovation and change has yet to be found. The difficulties in overcoming these forces are not to be underestimated.

Management Development, if it is to be an effective partner in Innovation, should encourage managers to be:

- outward looking rather than insular.
- questioning rather than passive.
- capable of understanding the "big picture" rather than parochial.
- confident of their own skills and abilities rather than afraid of being tested.

All these will help considerably to reduce resistance to change and make people more receptive to new ideas and new ways of doing things.

It is worth pausing to consider some of the characteristics of organisations which can inhibit change. One particularly notable inhibitor is strong inter-divisional or inter-functional rivalries which lead to the erection of strong boundaries which prevent the transfer of either ideas or people between parts of the organisation. Occasionally one of the major barriers is the separation (not always physical) of R&D departments from the mainstream of the organisation.

Both inter-divisional rivalries and the separation of R&D can be major inhibitors to the effectiveness of the Innovation process and the motivation of staff within the organisation. Where these barriers exist, a major task of top management should be to either reduce or eliminate the barriers by <u>positive</u> action. Management Development can assist in the process but the decision and the drive must come from the top.

Any organisation which wishes to undergo substantial change or Innovation, must ensure that the culture of the organisation is appropriate to the change or must put a significant amount of effort into managing the change in the organisation culture. Often, the major part of such a cultural change is a massive re-education and training programme from top to bottom of the organisation. However, it usually tends to be organisations that are in significant difficulties and are forced into the position of having to do something that actually tackle this problem in a constructive and positive manner. Successful organisations have a tendency to become a little complacent.

Organisation cultures that are supportive of Innovations of any description, tend to show the following characteristics:-

* intense communication especially on an informal level.
* toleration of failure and acceptance of risk.
* availability of resources for innovation and "flexible" control systems.
* long term time horizons.
* incentive and reward systems that encourage both Innovation attempts and successful Innovations.
* promotion of people who have a track record of Innovation is a key signal to the people in the organisation. If people who innovate or attempt to innovate, do not get promoted, then it should not be surprising if a climate of Innovation does not exist.
* Mobility of managers between Divisions and/or Functions.

No amount of Management Development activity <u>in isolation from</u> the commitment of top management and the total support of the organisation infrastructure can develop these characteristics in an organisation.

4. IMPLEMENTATION OF INNOVATION

In many cases, it seems as though managers responsible for implementing Innovation in the organisation are better qualified by both education and experience to develop the Innovation rather than manage its implementation. This is one area where Management Development can have a real and positive role to play in attempting to foster and develop implementation skills amongst those managers who will have to implement technological change. Many of these implementation skills are the same as the normal management of change skills that managers need. These can therefore be considered under the next heading.

What is important, however, is to identify which managers have the potential to move from working solely on the development of Innovations and move to implementing those Innovations and perhaps even managing the resultant manufacturing or operational process. If such an identification is made, then it is possible, by use of training courses, short secondments, and attachment to task forces etc, to develop the skills that will be necessary for managing the implementation process. Ideally, this process should start in advance of the need and can turn people who may, for example, be R&D specialists, into general managers.

There is one very important aspect that needs to be borne in mind for the implementation of technological Innovations; technological superiority, strategic importance and well documented potential benefits do not guarantee the acceptance of, and successful implementation of an Innovation.

5. MANAGEMENT OF CHANGE

There are several basic things which any manager who seeks to bring about change must be able to do. One of the best books on this area of change is "The Change Masters" by Rosabeth Moss Kanter "(ref.3)", from which these points have been adopted:

(a) Create a clear vision and sense of direction - The worst possible position to be in is one where there are no clear objectives and no clear purpose. I have already referred to this from the viewpoint of the organisation as a whole but for any manager attempting to implement change in a part of the organisation, the ability to create a clear vision and sense of direction that is appropriate for that particular part of the organisation or that particular task is crucial. Without a clear endpoint to aim at, no management team can unite around a goal.

It is worth mentioning that in this area, persistence is very important because there are many obstacles to success. Anyone embarking on major changes has to be robust enough to take setbacks and still keep going and as I have already mentioned, Innovations do not "sell themselves".

(b) Build coalitions - Technical Innovations do not implement themselves. People have to be convinced that they are beneficial. Support has to be built up, opposition neutralised or converted. This area obviously requires a high degree of political acumen, political skills, ability to read an organisation and empathy with many different managers in order to smooth the path to implementation. This will often be needed even if the proposed Innovation has top level support within the organisation.

(c) Work Through Teams - If effort over and above the norm is required and in any case of implementing technical change, it certainly will be, then one of the skills that will be desperately needed is that of getting the best out of people. Most often, this is best done by creating teams, giving people a sense of involvement, motivating them and keeping them going. Management Development and especially management training, can do much to ensure that managers understand the need for these skills, know what they are and are given the opportunity to develop them. Increasing Management Development practitioners actually work with Teams to ensure effective performance.

(d) Make Everyone a Hero - Share credit and recognition. This is a logical extension of the teamwork idea. If success is due to a team effort, then the boss should not claim all the glory for himself.

Management Development can obviously help give Managers the appropriate skills and knowledge to assist the change process but the key starting point remains the clear vision and sense of direction.

CONCLUSION

What I hope I have succeeded in demonstrating, is that Management Development is very dependent on the context in which it has to be undertaken, it does not exist in a vacuum. It is imperative that the Management Development approach for any organisation is consistent with both the vision or strategy of the organisation and the culture of the organisation.

Management Development is no substitute for an Organisation Strategy aimed at Innovation and an Organisation culture that supports Innovation. If the strategy is appropriate and the culture both supports and rewards Innovation, then Management Development has an important role to play both in enhancing the effectiveness of the organisation and in equipping the managers with the necessary skills to make Innovation a success.

ACKNOWLEDGEMENTS

I would like to acknowledge the help and advice of my colleagues at the National Management Centre and also the helpful comments of Mr D Needham, Assistant Director, Engineering Research Station, British Gas plc. The views expressed remain my own and do not necessarily reflect the views of British Gas plc.

REFERENCES

1 C. Freeman, "The Economics of Industrial Innovation", Penguin Modern Economic Texts, (1974).

2 H. Igor Ansoff, "Implanting Strategic Management", Prentice Hall International, (1984).

3 R. M. Kanter, "The Change Masters", George Allen and Unwin, (1984).

Innovation: Adaptation and Growth, edited by R. Rothwell and J. Bessant
Elsevier Science Publishers B.V., Amsterdam, 1987 — Printed in The Netherlands

THE ROLE OF THIRD-PARTIES IN THE DIFFUSION OF INNOVATIONS: A SURVEY

S.J. MANTEL, JR.[1] and G. ROSEGGER[2]

[1]Department of Quantitative Analysis and Information Systems, University of Cincinnati, Cincinnati, OH 45221 (USA)

[2]Department of Economics, Case Western Reserve University, Cleveland, OH 44106 (USA)

SUMMARY

In this paper, we undertake an exploratory survey of the impact of outside agents (third parties) on the diffusion of innovations. Formal diffusion models typically involve only two parties -- the innovator (originator) and the population of potential adopters (users). Less formal extensions of these models have recognized the roles of other parties in the process. While the influence of the government as a third party, acting through legislation and regulation, has been treated extensively in the literature, much less is known about the effects of other third-party institutions on the direction and rate of technological advance. We establish a tentative typology of these institutions and propose some rationales for their existence and and methods of intervention. Anecdotal evidence on the networks of interest that develop around the diffusion of many innovations is meant to illustrate the usefulness of our analytical framework.

INTRODUCTION

In this paper, we undertake a survey of some of the ways in which outside agents (we call them "third parties," or TPs, for short) may influence the diffusion of innovations. We define <u>third parties</u> as persons or organizations that intervene, with varying intensity of influence, in the adoption decisions of others. The third-party role does not necessarily imply that these agents are disinterested and "objective" factors in the diffusion process.

Our goal is to present a typology of TPs, briefly to explore reasons for their existence, and to provide some -- at this point mostly anecdotal -- evidence for their significance. What emerges from our survey is not so much a set of generalizations, as a number of suggestions for further research. While the influence of government regulation, implemented by a host of administrative bodies, has been discussed extensively in the technical-change literature (e.g., ref. 1), and while the role of "change agents" has been recognized in the writings of anthropologists and sociologists (ref. 2), we find that, with a few notable exceptions, other types of TPs have gone largely unnoticed. Among the exceptions, the works of Barke (ref. 3) and Link and Tassey (ref. 4) served to focus our attention on some important aspects of outside-agency influence.

Our approach here is based on the notion that, just as an innovation undergoes modifications that greatly affect its diffusion path, the institutional environment in which diffusion takes place also evolves in response to the expected or realized effects of widespread adoption. As we hope to show, various third-party institutions form an important part of this environment. Therefore, we believe that an attempt at a systematic survey of TPs is of more than taxonomic interest. In a period when governmental agencies are increasingly called upon to define, assess, and regulate new technologies, as well as to insure against their perceived risks (ref. 5), such a survey may help to suggest the extent to which market-driven or voluntary institutions perform many of the same functions.

A TYPOLOGY OF THIRD-PARTY INSTITUTIONS

In Figure 1 we propose a classification in the form of typical, stylized relationships among parties that might influence the diffusion of a new product or process. In order to place the subjects of our survey in the proper framework, we include what we call "the standard diffusion model" and some of its elaborations, as well as a synoptic version of regulatory impacts on adoption decisions. We shall briefly discuss these latter two, before turning to our main topic.

The standard diffusion model

Formal models of diffusion generally are patterned after the contagion paradigm and involve the dynamic interaction of just two parties -- the innovator(s) and the population of potential adopters. Such models can be specified in a variety of ways (ref. 6), but they focus on the quantifiable determinants of the rate at which potential adopters become actual adopters. A difficulty with these models is that they can account awkwardly, at best, for the transformations that "the innovation" is likely to undergo in the process of diffusion and which, in fact, continuously redefines the population of potential adopters (refs. 7-8). Therefore, empirical investigations and case studies often elaborate on the standard model's somewhat austere structure.

In these extensions, the accumulation of *information* is most frequently accorded a key role. Two such extensions are of interest in the context of our survey. The first considers the possibility of a kind of role reversal, in which early adopters not only participate in an innovation's initial development but are also active in marketing it, thus contributing to its more rapid diffusion (ref. 9-10). Voss, Chiaromonte, and others (refs. 11-13) found this information supplied by early adopters to be very important in explaining how new products "take-off" in the market. In a series of well-documented studies,

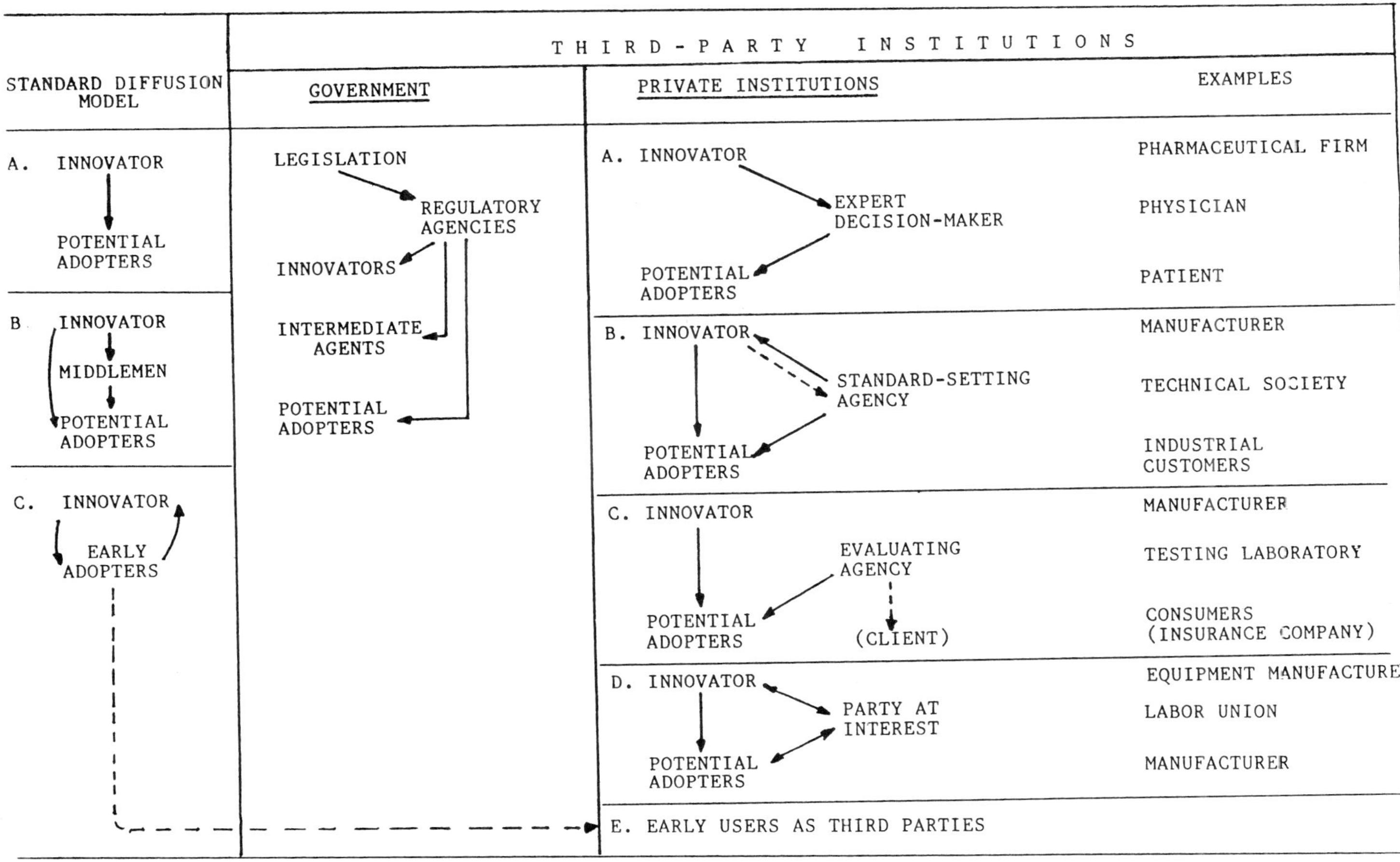

FIG. 1. TYPOLOGY OF THIRD-PARTY INSTITUTIONS

von Hippel showed that in a number of industries, innovation is characterized by a process wherein "users" developed new products and then sought manufacturers to produce them (refs. 14-15 and elsewhere). Although in our taxonomy we assign early users to the standard diffusion model, their role in the diffusion process shades over into that of a third party, and we take account of this fact in a subsequent section.

A second elaboration involves recognition of the intervention of middlemen, such as wholesalers and distributors, as well as of component suppliers, in the relationships between innovators and adopters. The concerns of these parties are, of course, readily defined in terms of the standard model, and therefore we shall not consider them further.

For our purposes, the important point is that early diffusion, at least of major innovations, quickly tends to create a network of economic, political, and psychological interests that influence and are influenced by the decisions of third-party institutions, such as specification-writing agencies. Nowhere is this as obvious as in cases where "voluntary" specifications, once agreed upon, are backed by the force of a government mandate (ref. 3).

Government as a third party

Obviously, administrative agencies play a key role in explaining the diffusion of "regulated" innovations. Whenever the public interest is at stake, mandates may lead to virtually instantaneous adoption of a product (e.g., seat belts, catalytic converters), or they may block or retard diffusion (e.g., the "anti-cancer drug", laetrile). Reversals and modifications of regulations frequently reflect changes in a technology or in the nature of information about it (ref. 16).

The so-called "government procurement specifications" are not, for the most part, compulsory. Rather they are a set standards for those producers who want to become vendors to various government agencies. Although in this case the government functions as a user and not as a third party, many of these specifications take on the force of a mandate, if only because innovators are rarely willing to ignore an important potential customer. This point becomes trivially obvious in the case of specifications promulgated by the Department of Defense.

Much less direct is the influence of such agencies as the General Accounting Office and the Office of Technology Assessment, but there can be little doubt that their pronouncements do affect both innovators and those contemplating adoption of an innovation (ref. 17).

Private third-party institutions

In the main part of Figure 1 we show several kinds of TP intervention that we consider prototypical for the role of private (voluntary), outside agencies in adoption decisions. Here we want to describe them briefly, reserving for the next section, an exploration of the rationales for intervention.

(i) The TP as an expert decision-maker. In many situations, the TP knows things about an innovation that the potential user does not know. In deferring to the TP's expertise, the user acknowledges that he or she regards the cost of personally acquiring such information as too high. This is certainly true for cases in which consultants exert a major influence on the diffusion of a new product. In other cases, reliance on an expert decision-maker may be mandated by law. The most obvious example is the requirement that drugs be prescribed by a licensed physician. The patient is presumed to possess enough information to choose among experts, but no more. Not too surprisingly, innovator-vendors direct their diffusion efforts at these various groups of third-party experts, not at the potential adopters themselves.

(ii) The TP as standard-setter/specification-writer. Standards or specifications, established by many different types of organizations, provide an *ex ante* framework for the evaluation of innovations. As constituted at any given time, standards provide information to innovators, qualify or disqualify new products and processes, or determine the parameters of their application by users. Conversely, innovations may give rise to pressures for revision of existing standards, whenever these inhibit diffusion.

(iii) The TP as an evaluator. When third parties conduct evaluations of, and provide information about, innovations *after* their introduction, these agencies will exert varying degrees of influence on diffusion. At one end of the spectrum is an old American institution, the Underwriters' Laboratories, established in 1912 by the insurance industry, whose tests of new products are crucial in determining market acceptance. At the other end, many of the more recently formed consumer organizations no doubt have influence on some individuals' decisions concerning the purchase of new products, but many such agencies have a rather limited circle of influence. Similar roles are also played by special-interest publications, aimed at firms and consumers, in many different fields of technology.

(iv) The TP as a party directly at interest. Whenever an innovation poses *ad hoc* advantages for, or threats to, the direct economic, social, political, religious (or other even less well defined) interests of particular groups, incipient diffusion will force these groups into action. The attitudes of labor unions toward technological advances have been explored in great detail and need no elaboration, nor is it necessary here to discuss the influence of special coalitions that frequently arise in response to the perceived benefits or costs of a specific innovation, such as nuclear power.

ROLES, RATIONALES, AND IMPLICATIONS

Our list of voluntary third-party types could easily be extended in several directions; however, it provides us with a sufficient basis for asking the question, "What are the rationales, economic or other, for the existence of these institutions and what are the implications for one of our society's avowed goals, the advancement of technology?"

The role of expert information

The expert, whether decision-maker, consultant, or early user, is a prime provider of information about innovations. The expert's role is defined by the precise nature of the relationship between the expert and potential adopters.

(i) The decision-maker. The archetype here is the physician. Because the doctor's permission is required for the patient to have access to most pharmaceutical products, the physician may act as a barrier to, or as an accelerator of, diffusion. The same applies equally to innovative diagnostic or treatment modalities. Intervening between producer and patient, physicians derive their -- legally modulated -- role from the presumption that they have more information about relevant characteristics of medical innovations than anyone else.

The physician's information, however, comes in the first instance from producer-innovators and only subsequently from experience, his/her own or that of other physicians. In light of the obvious possibility of bias in such information, and the potential for irremediable harm, a governmental agency (the Food and Drug Administration) intervenes to prescribe and standardize the information provided by producers.

Yet another set of TPs further complicates this relationship. Health insurance firms decide whether or not to include specific drugs or treatment technologies in their coverage, thus further circumscribing the physician's reliance on his/her presumed expertise. Whatever their particular interests, health insurance firms also become *de facto* influences on the diffusion of medical innovations. Finally, the accumulation of legal decisions in malpractice and liability cases has turned the courts into another "third party," whose effect on supposedly expert decisions is generally acknowledged, even if it is difficult to assess.

Clearly, these forces interact in different ways in specific situations, some accelerating and others retarding diffusion. The only generalization possible is that these interactions impose additional costs on all participants in the process. The complexities of such situations have been dealt with revealingly in Coleman *et al*. (ref. 18), and media attention has focused on the physicians' frustration over the steady reduction of their freedom in exercising their presumed expertise (ref. 19).

We concentrated on this, admittedly anecdotal, portrayal of an expert TP's role mainly to illustrate some of the difficulties in capturing all relevant factors in seemingly simple situations.

(ii) *The consultant*. The obvious distinction between the role of the decision maker and that of the consultant is that the expertise of the latter is not backed by any legal sanction. In practice, however, this distinction is not as sharp as it might appear. For one thing, the patient of our earlier example can always purchase but not actually take a drug prescribed by a physician, and for another, consultants often have behind them not only their technical expertise but also the force of information about legal and regulatory constraints, such as building codes or the rules of the Environmental Protection Agency.

The influence of the consultant is further enhanced in all situations where the consultant assists the adopter in implementing expert judgments. Thus, the computer consultant may, for example, design or purchase, and install hardware and software for a client. In such cases, the interests of suppliers and subcontractors once again complicate what, on the face of it, may seem like a simple and straightforward relationship.

Whenever several individual consultants or consulting firms have to be relied upon by adopters for various components of a complex innovation, there also exists the potential for conflicting advice. Although we may assume that potential adopters would not employ consultants unless they are favorably disposed toward the possibility of adopting an innovation, such conflicts may actually retard diffusion.

Nevertheless, the economic rationale for the consultant as a third party seems clear: (1) reliance on a consultant tends to lower information costs for potential adopters; (2) it tends to lower transaction costs in cases where adoption involves the acquisition of complex systems of equipment; and (3) having relied on the expertise of consultants, the adopter may have protected himself, at least partially, against the possible legal consequences of a decision made without the force of established expertise behind it. Social and psychological rationales for the use of consultants also exist. The use of well-known consultants to assist with the adoption of innovative managerial techniques (e.g., quality circles or MRP) can take on a fad-like fervor out of proportion with reasonable economic expectations.

The role of standard-setting agencies

The United States is unique among major industrial nations in that responsibility for establishing standards and specifications is commonly assigned to private, cooperative groups such as technical and professional societies and trade associations. The government intervenes by setting mandatory standards only in cases where the public interest is directly affected. But government also affects technological decisions through the specifications set for its own purchases. As we shall discuss below, however, there is considerable scope for interaction between voluntary organizations and various government agencies, with more or less desirable effects on the diffusion of innovations.

Approximately 400 organizations, ranging from the Aluminum Association to the Zinc Institute, have promulgated more than 20,000 standards and specifications for capital equipment, production methods, and final products. The distinctive feature of these associations in that compliance with their standards is voluntary. Nevertheless, the static-efficiency gains from standardization virtually assure their general acceptance. Link and Tassey (ref. 4) demonstrate that standards provide the benefits of information, compatibility, variety reduction, and quality assurance. These benefits certainly constitute a persuasive rationale for the role of standard-setting agencies in a technological system characterized by large numbers of complex interdependencies. We suggest, however, that in the dynamic setting of diffusion, additional considerations come into play that give rise to a somewhat less sanguine assessment.

In the typical association, committees of volunteer technical experts establish, update, and revise standards. Of necessity, most of these experts are employed by firms directly affected. In addition, associations differ widely in size of membership, resources, and administrative support. Some, particularly the small and understaffed associations, may promulgate standards and specifications that lag well behind technological developments and may revise them only infrequently. This will surely affect the diffusion rates of innovations. Other associations, with ample resources and solid technical support staffs, produce a steady flow of newly created and updated standards, in keeping with the latest technical advances in their fields. Thus, for example, the Society of Automotive Engineers (SAE) publishes annual, multi-volume sets covering thousands of process and product specifications. Mere size and capability do not, however, assure receptiveness for innovations. Even, or perhaps especially, large organizations may be dominated by the interest of their most influential members, and these interests may not lie in the rapid acceptance of major technological innovations. Only investigations dealing with specific industries and technologies could provide some balanced assessments of the relationship between standard setting TPs and the diffusion of innovations.

We speculate that, at least, the following factors will play a role:

(1) Where "systems integrity" requires complete standardization, such as in telecommunications, the diffusion of many innovations is essentially an all-or-nothing proposition. Standard setting obviously will be greatly influenced by those who have a stake in particular technologies, and diffusion may be delayed until all conflicts have been resolved. Barke's case study of the struggles surrounding adoption of a standard technology for color television (ref. 3) provides a good example. It also demonstrates that participation of government agencies does little to speed up the process.

(2) Existing standards can be so firmly entrenched that the spread of innovations appears altogether blocked. Every potential user sees the costs of modifying existing designs, equipment, or procedures as exceeding the otherwise unarguable benefits of an innovation. Paul David's study of the persistence of the obviously inefficient arrangement of the standard typewriter keyboard is a delightful example (ref. 20). Some seemingly innocuous standards, such as the dimensions of essential components (e.g., automotive spark plugs), may, in fact, be most difficult to change.

(3) Initially, major innovations often appear in several, functionally equivalent versions. Even without the strictures imposed by systems integrity, there will be obvious gains in making one of these versions the standard. The best contemporary example is provided by the repeated calls for greater "system compatibility" among the products of competing computer manufacturers. To the extent that it reduces uncertainty among potential adopters, early standardization of an innovation's essential parameters will, no doubt, hasten diffusion. On the other hand, early standardization may remove incentives for further innovative effort, thus freezing technology for prolonged periods. In spite of this, it is clearly futile to speculate about the optimal timing for the establishment of general standards or even about whether past standard setting for a particular technology occurred too early or too late -- especially since one does not know what directions technical development might have taken in the absence of standards.

(4) Even ostensibly voluntary standards may be used to limit competition among producers and to reduce the number of technical alternatives open to potential adopters. Ample evidence on this topic is provided by the industrial organization literature and by a host of recent studies of non-tariff barriers. We need not elaborate on these.

The role of evaluators

In principle, it is possible to classify the roles of third-party evaluators in terms of their concern with (a) those product or process characteristics of direct interest to the potential adopters (e.g., the Ralph Nader group's anti-Corvair campaign), (b) the actual or potential external effects of a widespread adoption of some new product or process (ref. 21), and (c) the political or economic interests of some specific groups other than innovators and adopters.

In practice, such neat distinctions frequently break down. Economic interest often masquerades as concern for the general good, rivalry among innovators is played out in the public arena, and risk assessments are colored by a desire to cater to irrational or sensationalist sentiments. Here again, the sorting out of forces tending to retard diffusion from those helping to accelerate the acceptance of innovations would be a matter for detailed case studies. While the issue of externalities, for example, has received considerable attention in the literature of technological change, we are not aware of similarly focused investigations of the impact of such TPs as the Underwriters' Laboratories or Consumers' Union. Similarly, the effects of court decisions on the evaluation of product liability risks by third parties have been discussed widely in popular and professional publications, but less appears to be known about the influence of what de Sola Price (ref. 22) has called the "boasting and heroics" about innovations in much of the technical literature.

That new information can lead to *ex post* evaluations that retard or stop the further spread of innovations that have already been widely diffused just adds another complication to the picture. The case of DDT provides an outstanding example (refs. 23-24).

Early users in the role of third parties

The phenomenon of early adopters of an innovation stepping into the role of third-party promoters has been observed. What motivates these TPs to take an interest in development and faster diffusion, instead of reaping the quasi-rents presumably accruing to them as a result of their pioneering? Our own, as well as other studies suggest that motivation comes from the realization that widespread testing, modifying, and marketing of the innovation will, in fact, increase their benefits in the longer run.

E. von Hippel's investigations (ref. 25), based on the "customer active paradigm," provide one set of illustrations. But the relationships between innovators and adopters are more varied than is suggested by the descriptors "producer active" or "customer active." Examining the development and spread of innovations in information technology, Voss (refs. 26-27) developed and tested a five-part classification of these relationships, organized around the roles each party played in initiating, innovating, developing, and marketing application software.

In several case studies, Mantel and Meredith (refs. 3 and 28) found levels of user involvement in the development, modification, and distribution of innovations that went well beyond what one might expect from firms operating at arm's length. Early adopters put on demonstrations, spoke to potential user groups, wrote articles for the trade press, and even acted as unofficial members of the innovator's sales force. The pay-off for their efforts as "temporary third parties" not only came in the form of large amounts of additional information, but also lowered psychological stress through risk-sharing, as well as the prestige -- and associated economic benefits -- of gaining a reputation as an industry leader.

We conclude, in any event, that early adopters, whenever they involved themselves in the role of third-party promoters, always contribute to the more rapid diffusion of innovations. No such unequivocal statement could be made about any of the other third parties.

CONCLUDING OBSERVATIONS

Our brief survey of third-party influences on the diffusion of innovations permits no sweeping generalizations. We undertook it because our firm-level studies of the innovation process had convinced us of the phenomenon's importance in explaining many types of otherwise puzzling behavior; however, our search of the literature showed that there existed no systematic treatment of the subject. Therefore, organizing some rather disparate material seemed like a useful first step toward further research. We hope that our typology, still tentative, will trigger suggestions for improvement, and we expect eventually to replace our casual and anecdotal evidence with more comprehensive data.

At this point in our work, we venture only one conclusion. There is room for much further exploration by students interested in the economic and managerial aspects of the diffusion process. We believe that a better understanding of third-party roles in this process will add substantially to our knowledge of how institutional arrangements in a market economy affect the direction and rate of technological advance.

REFERENCES

1 W.M. Capron, (Ed.), Technological Change in Regulated Industries, Brookings, Washington, 1971.
2 E.M. Rogers, Diffusion of Innovations, 3rd edn., Free Press, New York, 1983.
3 R. Barke, Regulation and cooperation among firms in technical standard-setting, Journal of Behavioral Economics, 14(Winter 1985).
4 A.N. Link and G. Tassey, The impact of standards on technology-based industries: The case of numerically-controlled machine tools in automated batch manufacturing, Working Paper EC080701, Center for Applied Research, Univ. of North Carolina at Greensboro.

5 Y. Aharoni, The No-Risk Society, Chatham House, Chatham, NJ, 1981.
6 P. Stoneman, The Economic Analysis of Technological Change, Oxford Univ. Press, Oxford, 1983.
7 G. Rosegger, Diffusion research in the industrial setting: Some conceptual clarifications, Technological Forecasting and Social Change, 9(1976) 401-410.
8 G. Rosegger, Diffusion and scale dynamics: A case study, Technovation 1(1982) 291-304.
9 G. Foxall, A conceptual extension of the customer-active paradigm, Technovation, 4(1986) 17-27.
10 S.J. Mantel, Jr. and J.R. Meredith, The user's role in spreading innovation: A micro view of technological diffusion, Atlantic Economic Conference, St. Thomas, V.I., April 15-19, 1986.
11 C.A. Voss, The management of new manufacturing technology: Eight propositions, Systemes De Production, APII (1985).
12 F. Chiaromonte, The strategic role of management of the innovation process, in: H. Hubner (Ed.), The Art and Science of Innovation Management, Elsevier, Amsterdam, 1986.
13 M.L. Markus, Power, politics, and MIS implementation, Communications of ACM, 26(6) (1983).
14 E. von Hippel, The dominant role of users in the scientific instrument innovation process, Research Policy, 5(3) (1976) 212-239.
15 E. von Hippel, Transferring process equipment from user-innovators to equipment manufacturing firms, R&D Management, 8(1) (1977) 13-22.
16 W.S. Peirce, Bureaucratic Failure and Public Expenditure, Academic Press, New York, 1981.
17 W.A. Hetzner, L.G. Tornatzky and K.J. Klein, Manufacturing Technology in the 1980's: A survey of federal programs and practices, Management Science, 29(8) (1983) 951-961.
18 J.S. Coleman, E. Katz and H. Menzel, Medical Innovation: A Diffusion Study, Bobbs-Merrill, Indianapolis, 1966.
19 S. Steptoe, Hassles and red tape destroy joy of the job for many physicians, Wall Street Journal, 68(125) (April 10, 1987) 1 & 12.
20 P. David, Clio and the economics of QWERTY, American Economic Review, Papers and Proceedings, 75(2) (1985).
21 M. Sun, Weighing the social costs of innovation, Science, 223(1984) 1368-69.
22 D.J. deS. Price, The structure of publication in science and technology, in: W.H. Gruber and D.G. Marquis (Eds.), Factors in The Transfer of Technology, M.I.T. Press, Cambridge, MA., 1969.
23 G.M. Woodwell, Broken eggshells, Science 84, (November 1984).
24 P. Drucker, New technology: Predicting its impact is perilous and futile, New York Times, (April 8, 1973).
25 E. Von Hippel, Get new products from customers, Harvard Business Review, 60(2) (1982) 117-122.
26 C.A. Voss, The innovation process in information technology, Journal of Product Innovation Management, 2 (1985).
27 C.A. Voss, A typology of user and supplier involvement in the innovation process, unpublished working paper, personal communication (1984).
28 S.J. Mantel, Jr. and J.R. Meredith, The role of customer cooperation in the development, marketing, and implementation of innovations, in: H. Hubner (Ed.), The Art and Science of Innovation Management, Elsevier, Amsterdam, 1986.

Innovation: Adaptation and Growth, edited by R. Rothwell and J. Bessant
Elsevier Science Publishers B.V., Amsterdam, 1987 — Printed in The Netherlands

MANAGING INNOVATION IN THE INTERORGANIZATIONAL INFORMATION ERA

Nanette S. Levinson

Associate Professor, Department of Computer Science & Information Systems

Hurst 206, The American University, Washington, D.C. 20016

ABSTRACT

New developments in information technology make possible information technology-based linkages among more than one organization. This paper presents a conceptual framework for examining such linkages and an analysis of a 1986 study dealing with information technology strategy and structure. The study covers thirty-two large U.S. organizations, primarily in the private sector. Findings from this study as well as from the literature reviewed possess key implications for successful innovation management in the interorganizational information era-- an era marked by information systems linkages within and across organizations.

Today's managers operate in a new environment impacted dramatically by rapidly changing technologies. Information and telecommunications technologies affect not only the ways in which managers carry out their tasks but also the industries themselves that set the background for these tasks [2]. Additionally, these technologies constitute the foundation for what I call the interorganizational information era-- an era wherein new patterns of information flow occur within sets of organizations involved in resource exchange relationships.

Managers who are particularly concerned with innovation increasingly need to focus on key linkages both within their own organizations and between their own organizations and significantly related outside organizations [8]. These linkages are being transformed as a part of the interorganizational information era. This paper presents a conceptual framework for viewing innovation management in the interorganizational era, reports on the relevant results of a thirty-two organization study, and discusses implications for effective management of innovation in the nineteen nineties.

CONCEPTUAL FRAMEWORK

An appropriate framework for examining innovation and information and telecommunication technologies in the interorganizational information era combines interorganizational theory, information transfer theory, and systems analysis and design theory. Interorganizational theory focuses attention on the exchange of resources among organizations connected by at least some dependencies or other relationships [3]. Focusing on a group of organizations related by resource exchange allows for an examination of an organization set' [5].

An organization set circa 1987 is a network of organizations related through significant resource exchange and influence relationships. For example, an organization set for a pharmaceutical company includes its suppliers (including technology suppliers), customers, competitors, research universities and government regulatory agencies. The pharmaceutical company serves as the focal organization within the organization set. (Defining the relationships within the organization set with reference to the focal organization allows for clear boundary-setting rules as well as for an interorganizational unit of analysis.)

Once an organization set delineates its boundaries (what's included in the set and what is outside the boundary of the set), information transfer theories become crucial. Information transfer theories highlight the flow of information across organizational and interorganizational linkages. The characterization of the linkages among the units in an interorganizational set is key in examining information flow. Characterization includes not only the mode and pattern of linkages but also the intensity of linkages [8].

Studies of boundary role personnel and communication patterns in R & D settings reveal much about the mode and pattern of linkages present when successful innovation management takes place. These studies indicate the presence of information flow among boundary role personnel linking the focal organization and key project participants with other significant organizations in an organization set [1, 13].

Most studies of information linkages focus on boundary personnel and their oral and written communications. With the advent of new information and telecommunications technologies, the potential for impacting existing linkages and creating new or different linkages is great.

Here systems analysis and design theory comes into play. Traditionally systems analysis and design theory focuses on intraorganizational concerns [12, 14]. Definition of functional requirements involves in-depth analysis of user needs for a new system whether it be an accounting or a marketing system as well as concomitant analysis of organizational implications. In addition to the more traditional structured approaches to the systems development life cycle, iterative development or prototyping approaches also adopt an intraorganizational stance. While these newer approaches provide a more active and involved role for users, their implementation still is rooted in an organizational rather than an interorganizational perspective.

Recent work in the areas of interorganizational theory and knowledge transfer theory possesses important implications for traditional systems analysis and design theory with its intraorganizational focus. Systems analysis and design theory provides the foundation for the design and implementation of systems within organization. The new awareness of the potential competitive edge of systems [2] calls for a new dimension in systems analysis and design-- an interorganizational dimension. This new dimension becomes particularly significant in the innovation management milieau where linkages across organizations are becoming more and more important [4,8].

As John Rydz points out, "closing the wheel's loops is the essence of innovation [11, p.60]." His 'wheel' includes customers, futures R&D, marketing, engineering, manufacturing, and service; it constitutes another way of highlighting linkages that are not merely intraorganizational. He warns managers against using data as if it were a baton being passed along in a relay race.

Recognizing the advantages of new technology in the interorganizational era allows for maximizing the connections among the spokes in the wheel and, indeed, in an organization set. The analysis and design of information systems rooted in an interorganizational perspective ensures a non relay race treatment of information flow in an organization set.

STUDY OVERVIEW

The 1986 study of top information systems management in thirty-two United States organizations included twenty-eight major companies from a variety of industries and four miscellaneous organizations. Field researchers conducted interviews with the highest level information systems manager in each organization; the interviews lasted approximately two hours and included an interorganizational perspective [9]. Data from the interviews was coded, analyzed, and organized.

DISCUSSION

The Context

In order to utilize information technology to its fullest in effective management of innovation, several developments in the management of information technology itself provide an important foundation. The first development is the emergence of the role of chief information officer [9]. This position moves the reporting relationship and structural home of information systems management from a Director of Management Information Systems reporting at least several levels down in an organization to a Vice President for Information Systems or the equivalent reporting directly to the President. (It is interesting to note another similar trend, the emergence of a chief technical officer also reporting directly to the President.)

Thirty-one per cent of the sample of thirty two organizations have their highest level information manager reporting directly to the President. An additional twenty-five per cent of the thirty-two large organizations have their highest level information systems manager reporting to an Executive or Senior Vice President of the organization. In almost all of these organizations, this high level reporting relationship is a relatively new one. This finding reflects the growing importance of a strategic orientation for information management.

No longer is information systems management a function that provides support for primarily fiscal functions. It now encompasses in a growing number of organizations: traditional management information systems, office automation, telecommunications, printing and publishing. Integrating these functions and broadening their focus to include interorganizational concerns sets the scene for a strategic role for information systems management.

This strategic role requires two key changes in information systems management role definition. The first is a change from viewing information systems and technologies as support tools to viewing them as strategic weapons themselves. And the second is a concomitant change from focusing primarily on functions within the organization to focusing on linkages between the organization and other key organizations.

In order to make these changes effectively and implement the new role definitions of today's information systems managers, the top information systems manager needs to report to the top executive in the organization. The top executive has the final say in strategy formulation and implementation. Clearly the top level information systems manager must occupy an organizational rank that is appropriate for strategy leadership in the organization. He or she can then interface regularly with the vice president for R&D as well as other vice presidents to ensure that information systems and technologies work hand in hand with R&D and other key functions. In this way they can provide a solid boost for innovation and a competitive edge for the organization.

One area of great variety is the actual title given to the top information systems manager. The top information systems managers in the thirty-two organizations have a wide range of titles. These include: Senior Vice President of Information Systems Development; Vice-President, Business Information Systems; Vice President of Information Services; or Director of Information Resource Management. This diversity in titles as well as the movement in reporting relationships reflects an absence of unanimity regarding the specific title and placement of the top information systems position in a very diverse group of organizations.

Yet there is a common thread that emerges from this study. In over half of those interviewed there is increasing emphasis upon a strategic location at or near the top of an organization for the information systems function of the late eighties and nineties. This is a position that provides a significant locus for the requisite synergy between information systems management and innovation management.

Information Systems Decision-Making and Innovation Managing

Differing patterns of information systems decision-making provide differing implications for innovation management. Again the sample of thirty-two organizations provides evidence of great variety. Twenty-two per cent of the organizations are decentralized when it comes to information systems decision-making. Here the Vice President for R&D, for instance, can decide on information technology policies and expenditures.

Oftentimes this Vice President works cooperatively with the top level information systems manager to develop a plan in support of R&D's goals. Additionally, in situations of decentralized decision-making, information systems personnel are often directly attached to organizational divisions. Thus, R&D would have its own information systems personnel as would other major units of the organization.

The remainder of the organizations in the study (except for four organizations where information was not available) have a centralized model of information systems decision-making or a variation thereof. A common, centralized model involves a top level unit charged with information systems policy and acquisitions decision-making. Often there is a high level steering committee consisting of the heads of the major divisions of the organization as well as the head of the information systems division that provides guidance and makes major information systems-related decisions.

The Vice-President for R&D usually sits as a member of the Steering Committee; this location and opportunity to work with counterparts representing manufacturing, marketing, sales, and finance provides a superb opportunity for a holistic approach to key linkages both within and outside the organization. This is particularly useful in implementing a competitive strategy requiring information technology-based linkages that create a synergy both between an organization's value chain activities and between these activities and key value chain activities of other significant organizations in an organization set.

The whole area of information systems planning and its relationship to an organization's strategic planning [6,9], is another important element in considering effective innovation management. Locating the information systems management function at the highest corporate levels allows for a full information systems-based role and perspective at a rank where participation with the top innovation manager in organization-wide strategic planning takes place.

In order to assess where an organization is in terms of these dimensions, whether or not the organization has a separate information systems plan and strategic plan needs to be determined. Then the absence or presence of a link (as well as the nature of the link) between the plans can be identified.

Thirteen per cent of the organization's studied have no strategic plan at the corporate level whatsoever. Twenty-five per cent have no MIS plan to assist in information systems decsion-making. As one information systems manager stated, we are doing "basic blocking and tackling" rather than looking at information technology as a competitive advantage.

Thus, the vast majority of organizations in this study have both an MIS plan and a strategic plan. Thirty-one per cent of the information systems managers interviewed report a strong linkage between their information systems plans and their strategic plans. They are activiely involved in the formulation of the corporate wide strategic plan. Another nine per cent of those interviewed report some linkage. Usually it is in the form of providing support for the strategic plan once the top managers in the organization identify their strategic objectives. One top information systems manager in this grouping reports that he has a "practical linkage that ensures the strategic plan can realistically be accomplished in conjunction with the MIS Plan." Thus, just under half of the organizations studied actually link their two planning processes, at least to some degree, and appear to recognize the strategic advantage of information systems in their organization. Viewing information technology itself as an innovation and linking it to an organization's strategic goals for innovation reinforces a corporate culture that promotes innovation.

Finally, the study provides data indicating the general information technology strategies of the organizations in the sample. Thirty-eight per cent of the thirty-two organizations have or are moving toward a centrally planned orientation. This orientation requires a central information systems management unit reporting to top management [10]. Although original definitions of a centrally planned strategy implied total centralization of information systems management, today's centrally planned strategies often combine centralization on some aspects of information systems management with decentralization of other aspects. For instance, one company has a centrally located information systems management unit responsible for setting standards, doing information systems planning, linking to the organization's strategic plan, and making information systems policy. Within these guidelines the remaining information systems functions are decentralized with information systems personnel assigned directly to specific operating units.

A centrally planned type of information technology strategy creates a useful foundation for effective innovation management. Having the central focus allows information technology to play a partnership role with R&D, manufacturing, and marketing policies and practices. It allows for connectivity and compatability across functions as well as across organizational boundaries. At the same time, the concomitant decentralization present within the strategic guidelines set by the central unit sets the scene for the flexibility needed in managing innovation.

The next largest proportion of organizations, thirty-four per cent, has a monopoly strategy. This strategy is based on the belief that information is a corporate good and that the organization must provide an internal group to control and manage information technologies [10]. A monopoly strategy does not require a strategic orientation. It lacks the divisional flexibility and high level focus requisite for a strategic use of information technology in innovation management.

Nineteen per cent of the thirty-two organizations use a free market strategy. Here top management does not need to be involved [10]. Divisions can either use internal information technology-related resources or external sources (Parsons). This strategy, while lacking the strategic orientation and top management focus of a centrally planned strategy, does provide for the flexibility necessary for the succesful management of innovation.

Six per cent of the organizations have strategies combined with a leading edge strategy. The leading edge strategy emphasizes information and telecommunications technologies R&D [10]; it is compatible with strategies needed for managing innovation. One organization in this grouping has a centrally planned strategy combined with leading edge; another has a free market strategy combined with leading edge.

Another three per cent of the organizations studied, has a scarce resource strategy that involves no planning and treats information technology as a scarce resource [10]. This viewpoint is becoming less and less common. It is rarely found in organizations concerned with innovation.

Finally, only three per cent of those organizations had in place an information-technology-based linkage between the organization and major customers. This aerospace industry organization reports an extremely costly linkage with major development testing sites-- a linkage that effectively assists innovation management.

CONCLUSION

The findings discussed above provide a basis for further studies. Widening the focus of investigation to include top innovation managers within an organization and comparing locations and role of these key individuals with their information technology counterparts is an area of special concern. Additionally, this sample of thirty-two organizations across a variety of industries only yields evidence of one organization's information technology based linkages between it and major customers. Another sample of organizations, this time stratified by industry, might include evidence of additional linkage patterns including information technology-based linkages within industry-university research arrangements and other innovative patterns.

Based on the study of thirty-two major U.S. organizations in a variety of industries, it is clear that there is still a very strong intraorganizational focus for information systems management. This intraorganizational focus now extends across functions rather than merely on disparate accounting, marketing, finance, and other functions. Yet in close to half of the sample, at least some elements of a strategic perspective are beginning to appear. These elements include new titles, top level reporting relationships, and organization set approaches. Organizations with these elements are poised to support innovative innovation management-- management that requires effective information linkages across internal functions and among organizations (especially customers) in an organization set.

REFERENCES

1. T.J. Allen, "Roles in Technical Communication Networks," Communication Among Scientisrs and Engineers, C. Nelsen and D. Pollock, eds., Lexington, Mass., D.C. Heath, Lexington Books, 1970, 191- 208.

2. E.K. Clemons and F.W. McFarlan, "Telecom: Hook Up or Out," Harvard Business Review, July-August 1986, 91-97.

3. K. Cook, "Exchange and Power: Networks of Interorganizational Exchanges," Sociological Quarterly, Winter 1977, 62-82.

4. P. Drucker, "The Discipline of Innovation, " Harvard Business Review, May-June 1985, 67-72.

5. W.M. Evan, Organization Theory, New York, Wiley, 1976.

6. J. C. Henderson and M. E. Treacy, "Managing End-User Computing for Competitive Advantage," Sloan Manasgement Review, Winter 1986, 3-14.

7. B. Hodge and J.P. Clements, Business Systems Analysis, Prentice-Hall, 1986.

8. N.S. Levinson & D.D. Moran, "R&D Management and Organizational Coupling," IEEE Transactions On Engineering Management, Vol. EM-34, Number 1, February 1987, 28-36.

9. N.S. Levinson and H. Holley, Information Technology Strategy and Structure, Technical Report, 1986.

10. G. Parsons, Fitting Information Systems Technology To The Corporate Needs: The Linking Strategy, Harvard Business School, 9-183-176, Cambridge, Mass., U.S., 1983.

11. J. S. Rydz, Managing Innovation, Ballinger Publishing, 1986.

12. R. H. Sprague, Information Systems Management In Practice, Prentice-Hall, 1986.

13. M. Tushman and W. Moore, eds. Readings In The Management of Innovation, New York, Pitman Press, 1982.

14. E. Yourdon, Managing The Systems Life Cycle, Yourdon Press, 1982.

Innovation: Adaptation and Growth, edited by R. Rothwell and J. Bessant
Elsevier Science Publishers B.V., Amsterdam, 1987 — Printed in The Netherlands

COOPERATIVE RESEARCH IN U.S. MANUFACTURING*

A.N. LINK AND L.L. BAUER

Economics Department, The University of North Carolina at Greensboro, Greensboro, NC 27412 (USA)

SUMMARY

The U.S. government's most recent policy initiative to increase industrial research activity is the National Cooperative Research Act of 1984. Since its passage, much attention has been given to this new organizational research form, but to date there has not been a systematic investigation of cooperative research activity. This paper attempts to fill that void.

Interfirm differences in cooperative research activity are examined using a survey-based sample of 92 R&D-active firms from the U.S. manufacturing sector. From this descriptive analysis we conclude that: (1) market power has a significant influence on a firm's decision to engage in cooperative research; (2) the rate of productivity growth in cooperative research-active firms is not statistically different from that of non-cooperative research-active firms; and (3) the economic motives for engaging in cooperative research are industry-specific.

INTRODUCTION

The U.S. government has a long history of supporting public and private sector R&D. The Department of the Navy's sponsored research programs can be traced as far back as 1789, and the Department of Agriculture's involvement in the land-grant college system dates from the mid-1800s. Since World War II, direct support for R&D has increased dramatically in response to growing military needs. In the early 1950s, the government began to rely on tax incentives to increase private sector R&D. Most recently, a 25 percent R&D tax credit was introduced as part of the Economic Recovery Tax Act of 1981 (and renewed at 20 percent in 1986). Also, in response to a continuing decline in the rate of productivity growth in the manufacturing sector, Congress passed the National Cooperative Research Act of 1984 (hereafter referred to as the Act).

The intent of the Act is to increase innovative activity in the private sector by removing antitrust constraints on cooperative research. The Act establishes a rule of reason for evaluating the antitrust implications of R&D joint ventures, and it limits potential liability to actual damages rather than treble damages. While this Act has been in effect for over two years, still very little is known about cooperative R&D in general, or about the

*This paper is based upon research sponsored by the National Science Foundation under grant PRA85-212664.

productivity growth implications of cooperative R&D in particular.

COOPERATIVE RESEARCH AMONG U.S. FIRMS

The Act requires that any party to a cooperative R&D venture file a description of that activity with both the U.S. Attorney General and the Federal Trade Commission. As of December 23, 1986, 52 filings had been reported in the Federal Register. It is difficult to generalize about the nature of firms who engage in cooperative research from these filings. Thus, to investigate in detail the character of cooperating firms, survey data were gathered.

A random sample of 436 firms, stratified by manufacturing industry, was selected from the "R&D Scoreboard" in Business Week to receive a mail survey related to cooperative research activity. Cooperative research was defined to be any formal or informal relationship among firms for the purpose of conducting research. This definition includes consortia arrangements, but not university-based research relationships. After an initial mailing in 1985, and a follow-up mailing in early 1986, data had been collected from 92 manufacturing firms regarding their involvement in cooperative research.

Of the 92 firms, 62 reported that they were involved to some degree in cooperative research during 1984. The data in Table 1 show that this group of 62 cooperative research-active firms had been involved in 446 cooperative research projects. And, these 62 firms allocated, on average, 7.3 percent of their R&D budget to cooperative research activities.

TABLE 1 Distribution of Cooperative Research-Active Firms by Industry

SIC Code	Industry	Sample Firms	Cooperating Firms	Number of Cooperative Projects	% of R&D to Cooperation
20	Food	6	5	22	5.8%
21	Tobacco	0	0	0	-
22	Textiles	1	1	6	2.6%
23	Apparel	0	0	0	-
24	Lumber	0	0	0	-
25	Furniture	0	0	0	-
26	Paper	4	2	43	2.3%
27	Printing	1	0	0	-
28	Chemicals	15	8	109	7.3%
29	Petroleum	7	7	65	6.3%
30	Rubber	1	1	9	5.5%
31	Leather	1	0	0	-
32	Stone, Clay, Glass	3	2	21	5.1%
33	Primary Metals	3	2	3	4.0%
34	Fabricated Metals	4	3	4	6.2%
35	Machinery	14	10	61	6.8%
36	Electrical Equipment	16	10	75	8.8%
37	Transportation	8	5	15	8.1%
38	Instruments	6	4	8	12.1%
39	Miscellaneous	2	2	5	13.8%
		92	62	446	7.3%

INTER-FIRM DIFFERENCES IN COOPERATIVE RESEARCH ACTIVITY

As seen from Table 1, there is considerable variation in cooperative research activity among the firms in our sample. Why are some firms in the sample involved in cooperative research, and others not? Why, among those that do engage in cooperative research, are some firms allocating a larger portion of their R&D budget to that activity than others?

Two economic factors were investigated as possible determinants of inter-firm differences (n=92) in the percentage of R&D allocated to cooperative research activity: profitability and market power.

Profitability

As is well known, the R&D-to-profits relationship is stronger in some types of firms than in others. Theoretical analyses conclude that marginal firms, those engaged in "risky" R&D ventures, face an active self-financing constraint, and thus current profits are a necessary condition for R&D spending. Established firms doing "routine" product development will not be as constrained by cash availability and will not require high current cash flows to finance their research programs (ref. 1). Related empirical research supports this generalization (ref. 2). Given this, and the fact that cooperative research is likely to be oriented toward the basic or generic end of the R&D spectrum (ref. 3), it follows that high profits should be a prerequisite for firms to be active in that form of research. Hence, profitability and involvement in cooperative research should be positively related.

However, a countervailing argument is possible. Firms with low or declining profits may feel pressure to innovate in order to remain competitive (ref. 4). If so, increased R&D is a reasonable method for pursuing such a strategy. And, because of waning profitability, cooperative research may be the preferred organizational arrangement for such undertakings owing to the ability of the participants to share costs. So, firm profitability and involvement in cooperative research should be inversely related to each other.

Therefore, the theoretical literature on the R&D-to-profits relationship is conflicting, although the empirical literature suggests a positive relationship between these two variables. It is an empirical question as to whether this same positive relationship also holds for research conducted cooperatively.

Market Power

One of the most frequently investigated topics in industrial organization is the relationship between market power and innovative activity. Specifically, is it the possession of, or rather the quest for, market power that stimulates innovation? The so-called Schumpeterian position is that market dominance is an important determinant of innovation. Since cooperative research has a greater degree of "publicness" than internal R&D, by design, firms who have the ability to internalize the benefits from shared knowledge are, following Schumpeter, more likely to invest in it than would other firms. However, others have demonstrated that the relationship between innovation (R&D investments) and monopoly power cannot be unambiguously predicted from theory. The empirical research is also mixed (ref. 5 & 6). Thus, the relationship between involvement in cooperative research and market power is also an empirical issue.

Analysis

The survey data described in Table 1 were used to test these issues. Participation in cooperative research was measured as the percentage of each firm's company-financed R&D allocated to that activity (CRPCT) in 1984. Profitability was measured as the ratio of 1984 profits to firm sales, as reported in Business Week. Market power was quantified (based on the survey data) for each firm in two ways: by a weighted industry concentration ratio (WCR) and by a weighted market share variable (WMS).

Owing to the fact that the dependent variable, CRPCT, is truncated at zero, the relevant Tobit model for estimation is:

$$y_i = \begin{cases} y_i^* & \text{for } y_i^* > 0 \\ y^* & \text{for } y_i \leq 0 \end{cases} \tag{1}$$

where y_i refers to the dependent variable CRPCT, $y_i^* = X_i\beta+\varepsilon_i, E(\varepsilon_i) = 0$, $Var(\varepsilon_i) = \sigma^2$, and where X_i is the vector containing independent variables PROFIT, WCR, WMS, and a constant. The estimated Tobit coefficients from equation (1), along with OLS coefficients for comparison, are reported in Table 2.

TABLE 2

Tobit and OLS Regression Estimates Based on Equation (1)
(ratio of $\hat{\beta}/s_{\hat{\beta}}$ in parentheses: n=92)

Independent Variables	Tobit	OLS
Profit	-0.207	-0.017
	(0.88)	(-0.11)
WCR	0.128	0.126
	(2.74)**	(3.69)**
WMS	0.199	0.178
	(2.02)*	(2.39)*
Constant	-0.256	-0.012
	(-1.12)	(-0.71)
Standard Error	0.075	0.056
R^2	-	0.229
F-ratio	-	8.71**
X^2_3	16.285**	-

**significant at .01 level
*significant at .05 level

Profitability does not appear to be an important determinant of firms' participation in cooperative research. While the estimated Tobit coefficient of PROFIT is negative, it is not significantly different from zero.

The market power variables are statistically important. Firms in more highly concentrated industries, and those with larger market shares, are more active in cooperative research. The Tobit coefficient on both variables is positive and significant at the .01 and .05 levels, respectively. This quantitative finding is not at odds with the interpretation of Schumpeter discussed above.

Twelve two-digit industry dummies were included in other versions of equation (1), but as a group these variables were not significant and thus deleted. As well, the possibility of non-linear relationships between WCR and WMS with CRPCT was tested, but the non-linear terms were not significant.

Thus, our analysis suggests that systematic differences in the extent of cooperative research activity across the firms in this sample reflect the firms' abilities to appropriate, through their market power, the knowledge gained from that form of research.

COOPERATIVE RESEARCH AND PRODUCTIVITY GROWTH

One factor that motivated the passage of the National Cooperative Research Act was the belief that it would lead to more industrial R&D activity, and more R&D would thus help reverse the declining trend in productivity growth that has characterized the U.S. industrial sector for the last 20 years (ref. 7).

To examine the relationship between cooperative research activity and productivity growth, we calculated the average annual rate of growth in total factor productivity between 1981 and 1984 for each of the 92 firms. Using firm data from Compustat on residually-measured productivity growth, $\dot{A}/A = (dA/dt)/A$ was calculated from:

$$Q = Ae^{\lambda t}K^{\alpha}L^{(1-\alpha)} \tag{2}$$

where λ is a disembodied rate of growth parameter and α is the output elasticity of capital, as:

$$\dot{A}/A = \dot{Q}/Q - \alpha\dot{K}/K - (1-\alpha)\dot{L}/L. \tag{3}$$

The mean average annual rate of productivity growth between 1981 and 1984 was calculated for each of the 92 firms: for the 62 cooperative research-active firms it is 0.00913 and for the 30 noncooperative research-active firms it is 0.00861. The two mean values are not statistically different from each other (t-statistic = 1.077). In other words, there is no statistical difference between the productivity growth rate of cooperative and non-cooperative research-active firms in this sample.

MOTIVES FOR PARTICIPATING IN COOPERATIVE RESEARCH

To investigate, in an exploratory fashion, motives for engaging in cooperative research, each firm's R&D vice president was asked to indicate the strategic objectives associated with their firms' involvement in cooperative programs. Specifically, they were asked to distribute the proportion of their R&D budget allocated to cooperative research (CRPCT) between the following three strategic objectives: horizontal diversification into new product lines (HDPCT), vertical integration (backward and forward) (VIPCT), and leap-frog competition within existing product lines (LFCPCT). For the 62 firms engaged in cooperative research, the majority indicated that their primary strategic objective was horizontal diversification into new product lines.

As previously shown in Table 1, the survey sample of 92 firms comes from a number of two-digit SIC industries, with some not being represented at all. The subsample of 62 cooperative research-active firms is even more dispersed; however, some industries are sufficiently represented in the subsample to allow for a preliminary inter-industry comparison.

Table 3 shows the distribution of 45 of the 62 cooperative research-active firms by two-digit SIC industry. These 45 firms fall into six different industries: all other SIC industries have less than 5 firms each. Also shown in Table 3 are the percentages of company-financed R&D allocated to cooperative research in general and to the three categories of use (CRPCT = HDPCT + VIPCT + LFCPCT).

TABLE 3

Distribution of Cooperative Research Activity by Strategic Objectives by Two-Digit SIC Industry

SIC Code	Industry	n	CRPCT	HDPCT	VIPCT	LFCPCT
20	Food	5	5.58%	3.22%	0.84%	1.52%
28	Chemicals	8	7.29%	3.31%	0.33%	3.65%
29	Petroleum	7	6.29%	3.43%	1.86%	1.00%
35	Machinery	10	6.76%	3.22%	1.69%	1.89%
36	Electrical Equipment	10	8.80%	2.70%	3.25%	2.85%
37	Transportation	5	8.10%	1.81%	4.59%	1.70%

Column headings are defined in the text.

The distribution of means, albeit from samples of relatively small sizes, is interesting. It appears that the strategic motive for engaging in cooperative research varies across two-digit industries, as evidenced by the relative distribution of R&D dollars allocated to each activity. In the food, petroleum, and machinery industries the dominant objective is horizontal diversification into new product lines. The dominant objective in the chemicals industry is leap-frog competition into existing product lines, and in the electrical equipment and transportation equipment industries it is vertical integration. The extent to which different strategic objectives affect the performance of firms' R&D (that is, overall returns to R&D, for example) was not investigated.

CONCLUSIONS

The principal conclusion from this study is that there are systematic differences in the extent of cooperative research across firms in the manufacturing sector. These differences reflect to a large degree firms' abilities to appropriate through their market power the innovation-related knowledge gained in that form of research. It is also important that firm differences in cooperation are not an industry-specific phenomenon: holding constant market power, involvement in cooperative research does not vary systematically across industries.

There is no quantitative evidence that firms involved in cooperative research have a higher rate of productivity growth than firms who are not cooperative research-active. This finding in itself should not be interpreted as an indication that the Act will be ineffective in stimulating innovation, but it does suggest that either more time is needed to evaluate fully its impact or that cooperating firms have yet to realize the potential benefits from research.

Finally, while the amount of resources allocated to cooperative R&D is not an industry-specific phenomenon, there is suggestive evidence that the strategic motives for engaging in cooperative research do differ significantly across industries. Only descriptive data in support of this conclusion are offered here.

These conclusions, even as general as they are, must be interpreted with caution. First, since this is the initial empirical investigation of the topic, there is no other information available for comparison. Public information in Federal Register filings appears to be of limited use for determining any general characteristics of firms who engage in cooperative research. Also, while the survey-based sample of 92 firms analyzed here accounts for a sizeable portion of total R&D in manufacturing, it is not necessarily representative of all innovative firms: R&D is only one source of innovation-enhancing knowledge. And finally, the empirical analyses are simplistic owing to data limitations.

Still, the conclusions offered are interesting and indicate a strong need for additional empirical work on this topic.

REFERENCES

1 M.I. Kamien and N.L. Schwartz, Self-financing of an R&D project, American Economic Review, 68 (1978) 252-261.

2 A.N. Link, An analysis of the composition of R&D spending, Southern Economic Journal, 49 (1982) 342-349.

3 B. Bozeman, A.N. Link and A. Zardkoohi, An economic analysis of R&D joint ventures, Managerial and Decision Economics, 7 (1986) 263-266.

4. Horowitz, Estimating changes in the research budget, Journal of Industrial Engineering, 12 (1961) 114-118.

5 R.C. Levin, W.M. Cohen and D.C. Mowery, R&D appropriability, opportunity, market structure: New evidence on some Schumpeterian hypotheses, American Economic Review, 75 (1985) 20-24.

6 J. Scott, Firm versus industry variability in R&D intensity, in: Z. Griliches (Ed.), R&D, Patents, and Productivity, U. of Chicago Press, Chicago, 1984, pp. 233-245.

7 A.N. Link, Technological Change and Productivity Growth, Harwood Academic Publishers, London, 1987.

Innovation: Adaptation and Growth, edited by R. Rothwell and J. Bessant
Elsevier Science Publishers B.V., Amsterdam, 1987 — Printed in The Netherlands

RESEARCH AND DEVELOPMENT AND COMPETITIVE PERFORMANCE IN BRITISH AND AMERICAN HIGH TECHNOLOGY SMALL FIRMS

R.P. OAKEY[1], R. ROTHWELL[2], M. BEESLEY[2] and S.Y. COOPER[1]

[1]Department of Business Organisation, Heriot-Watt University, 31-35 Grassmarket, Edinburgh, EH1 2HT, Scotland

[2]Science Policy Research Unit, University of Sussex, Mantell Building, Falmer, Brighton, BN1 9RF, England

SUMMARY

The generation of relevant technical information is widely acknowledged as a critical first step towards successful product and process innovation in small high technology firms. This information can be gained from internal R&D, external sources, or through a combination of these means of acquisition. This paper examines the methods of management used by small high technology firms in organising their technical information gathering activities, and compares management behaviour with subsequent economic performance.

THE SEARCH FOR TECHNICAL KNOWLEDGE

The acquisition of technical knowledge is a fundamentally important step in the development of the products and services that high technology small firms seek to sell in competitive industrial markets. While the acquisition of a theoretical "optimal" amount of necessary technical knowledge is a level of precision beyond the competence of most normal satisficing small firm managers, implying that in many cases small firms will either perform insufficient or excessive amounts of technical information gathering, all firms will need periodically to update their products and/or services and methods of manufacture in order to preserve a competitive edge. Thus, a major strategic problem facing small high technology firm management centres on an appropriate level of investment of their precious resources in the search for technical knowledge. In this sense, the problems of decision making concerning the acquisition of technical information differ little from other decisions made by manufacturing firms (refs. 1-4). However, uncertainties surrounding the search for technical information are compounded by the reality that, not only is the optimal amount of search required difficult to quantify, but also the benefit of the information obtained for a given cost cannot be predetermined in order to inform the advisability of initiating its acquisition.

Set in this context, the performance of internal research and development is merely one option in a whole range of possible strategies open to the high technology small manufacturing firm in its quest for new technical information. Since it is simplistic to assume that there is a perfect causal relationship between increasing research and development and subsequent success, measured in terms of sales and/or employment growth, the successful management of technical information acquisition will depend upon the achievement of the most appropriate level and form of technical information acquisition, given the needs of the firm, which are determined by its type of activity. For example, a sub-contract printed circuit board manufacturer will generate a less sophisticated range of technical requirements than a product-based firm developing a new instrument based on laser-optics. However, it is essential at this point to stress that the differing technical requirements of the two firms do not necessarily imply different potentials for growth and prosperity. There is an unjustifiable tendency in many journalistic academic writings on high technology industry to employ a form of "technology snobism" where technical complexity is judged as synonymous with economic success. The example of the Concord supersonic passenger airliner indicates the folly of such an approach.

The earlier work into high technology small firms on which this new study is based rightly stressed the critical importance of risk in determining the small high technology firm's choice of technical information acquisition strategy (ref. 5). While it is true that some firms within high technology industries based on relatively simple technologies may not need to invest in expensive full-time research and development, a full-time research team is often essential for the most technically sophisticated firms. However, a strong element of risk in the adoption of a full-time research and development staff in such firms is the high cost of salaries for such workers during the long "lead times" (perhaps three to five years) that are often associated with the development of new high technology products. It is clear that in many cases, the desirability of full-time research and development is tempered by the reality that this form of activity is the most costly, and subsequently the most risky form of technical information generation attempted by small high technology firms.

The combination of the variable applicability of research and development and problems associated with its cost and subsequent risks ensure that a wide range of internal R&D and external means of technical information acquisition may be employed by the high technology small firm. As a general principle, the higher quality technical information is usually more expensive to generate and is predominantly performed by a full-time internal research and development department staff. Figure 1 is a visual aid to an understanding of the major types of technical information gathering strategy open to small high technology firms.

The matrix in figure 1 exemplifies a range of technical information generation strategies that cater for increasing levels of both investment (and subsequent risk), and management control of output. In general terms, the quality and subsequent cost of generating novel technical information increases towards the bottom of figure 1.

The figure is useful in indicating the generally strong relationship between the quality of the information generated, its cost, and most importantly, the degree to which such information is internally generated and controlled. For it is significant to note that although the purchasing of a licence or the setting up of external contract R&D is costly, these options do not rate highly on the "internal management and control" dimension of figure 1. This dipiction is justified on two counts. First, although external contract research may be costly, the commitment of the small firm to medium or long term funding may be more limited (e.g. a fixed price contract), when compared with the levels of commitment and control implied by the employment of a full-time research and development staff. Second, externally generated research, regardless of its sophistication, cannot be built upon internally if there is little or no competence within the firm for the continuation of development work on technology obtained from external sources. Thus, the existence of a full-time research and development staff confirms the continuing ability of the firm to generate and independently control high quality confidential information, without the need to resort to external sources of technology. This internal independence is particularly important for sophisticated product development where close internal liaison between design and construction is important.

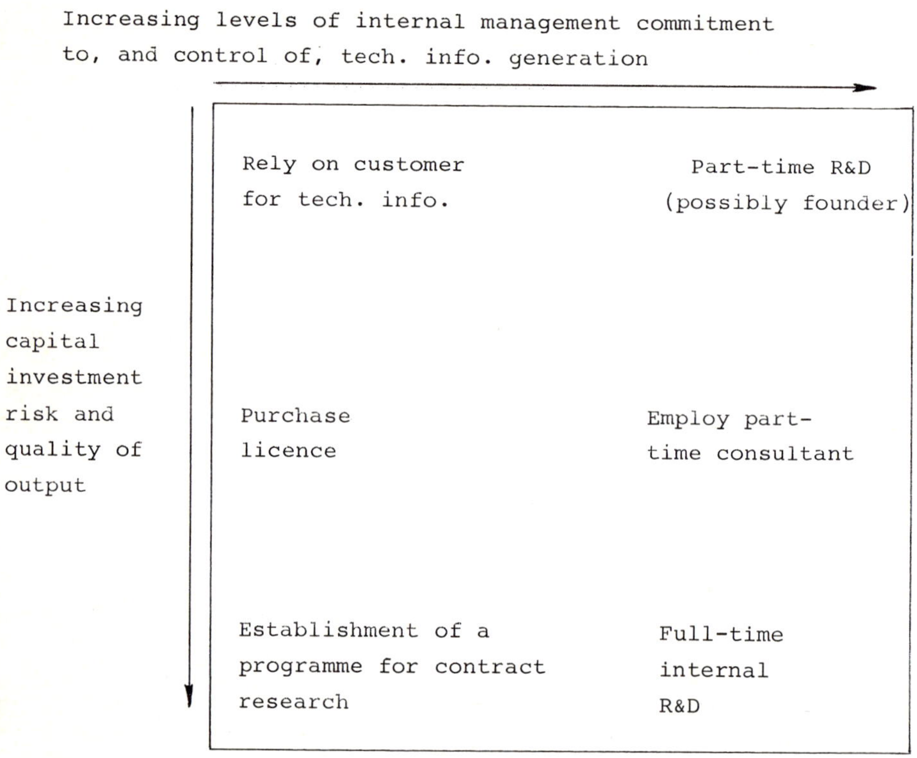

FIGURE 1
Technical Information Gathering Strategies

Other possible strategies of the high technology small firm include the purchasing of a licence, which usually implies that the acquired technology is not of the most desirable "leading edge" nature, and certainly not subject to full internal control. The employment of a retained consultant is another option which may be neither highly productive nor efficient in terms of innovation or internal small firm control.

Finally, reliance on customers for technical information again implies neither sophistication nor internal control of such technology (perhaps typified by a sub-contract printed circuit board maker). However, although the performance of part-time R&D, often by a technical entrepreneur, ensures internal control of the research, the piecemeal nature of such work frequently results in a poor volume and quality of output. Clearly, it is possible for the firm to utilise simultaneously a number of the options identified in figure 1. As implied above, the adoption of any of these strategies may ensure that firm's management has adequately fitted the form of technical information acquisition to the needs of the firm, radically exceeded, or undermet its "objective" requirements.

A further examination of internal research and development, risk and success in the small high technology firm.

The introductory discussions thus far have tended to suggest that the maintainance of an internal full-time research and development department is a risky but worthwhile "luxury" for the product-based small high technology firm engaged in cycles of product innovation which may be very short in fast moving areas of technological change. While this assertion probably holds as a "necessary evil" for the product-based firm, it is worthwhile emphasising in further detail at this point the basic risks with internal full-time research and development that render such activity particularly fraught with danger for the small high technology firm. These risks centre on <u>two</u> major uncertainties that the small firm management cannot eliminate as a programme of internal research and development begins.

i) Risks of escalating cost that eminate from the inability of small firm management to accurately predict the likely length of a research project. In particular, since high R&D wage costs are heavily dependent upon the duration of a project, which may burden the firm three to five years in advance of any return from sales, there is considerable scope in high technology firms for R&D costs to bankrupt the firm <u>before</u> the sales from a new product innovation can be enjoyed.

ii) A second risk surrounds the technological quality of the new product innovation. There is clear scope for the technological advantage afforded by an expensive programme of innovation to be insufficient to justify the resources invested in research and development. While few programmes of innovation are total failures, there is considerable scope for a poor return on a considerable investment in research and development.

In concluding this introduction, prior to the analytical sections, it is worth restating that the acquisition of technical information, either by internal or external means, is no guarantee of successful innovation and subsequent economic success. The above discussion has stressed that the key to good management of the technical innovation process in small high technology firms is the accurate matching of information generation to the particular quality and volume needs of the individual firm. This level of need will vary depending on the sophistication of the firm and may be met by a combination of both internal and external methods of generation. In the absence of relevant information, there is often a simplistic tendency to use the crude volume or quality of innovative activity as a surrogate measure of "success" in the absense of hard evidence to correlate information gathering behaviour with economic performance. Put simply, it is frequently assumed that evidence of "best practice" in terms of technical information generation (e.g. a large internal full-time research and development department) is a measure of "success". However, it is clear that all that has been accurately measured is "success" in terms of research and development itself, and is not a more meaningful measure of success judged in terms of subsequent employment, turnover and profit growth.

It may transpire that a firm with a small investment in technical information gathering has produced the most successful growth performance through the optimal matching of a modest level of research and development to a modest need, thus avoiding highly wasteful research and development expenditure which produces a poor benefit at a high cost.

This research is fortunate in that the two studies on which this work is based have provided the opportunity to construct a number of performance variables which will enable the analysis to compare hypothesised "best practice" behaviour with subsequent actual economic performance. Thus, there will be an opportunity to compare the economic performance of modest generators of technical information with their highly active counterparts. These investigations will be set in the broader context of examining the evolving management performance of three British and American regional sub-groups of firms. Such investigations may lead to evidence that takes the examination of the impact of research and development on small high technology firms beyond the simplistic "more R&D equals more success" approach to the subject of technical information acquisition.

INTERNAL RESEARCH AND DEVELOPMENT AND INNOVATION

The incidence of internal research and development

In advance of sections of this paper that include data from two surveys of high technology small firms, a number of basic statistics on the survey designs will assist an understanding of the following tables. The first study was performed between 1981 and 1982 in the geographically diverse industrial regions of Scotland and South East England in the United Kingdom, and the Bay Area of California in the United States (for full details of methodology - See ref. 5, chapter 5). These regions were chosen to give a full spectrum of industrial environments, ranging from the development region of Scotland, to the Bay Area of California, noted for its high technology small firm growth. This study took place through personal interviews in 174 instrument and electronics firms, divided almost equally between the three regions. The original study was designed to examine the impact of varying local external environments on internal innovation. The results of this work have been reported in a recent book (ref. 5).

The "time series" data included in this paper result from a follow-up survey in the 131 firms that remained for interview from the 174 firm 1981-82 survey in 1985-86 (which was reduced by 31 closures and 12 non-respondents). Personal interviews were conducted in 48 firms (i.e. sixteen per region), while the balance of firms in each region were surveyed by a less detailed telephone survey which duplicated key questions within the personal interview schedule. This most recent study is aimed at investigating the internal management behaviour towards innovation and its regional variations. As will be readily noted from the following data, the five year "time-series" statistics on the performance of these survey firms permits hard evidence against which often subjective personal evaluations of economic performance and management success may be gainfully compared.

While it is technically feasible for a high technology firm to maintain a high level of innovation by the use of totally external technical information sources, most of the available evidence from earlier work on this data set (ref. 5), and other work into innovation in small high technology firms (ref. 6), suggest that, in most cases, some form of internal research and development capacity is essential. Much of the informal innovative atmosphere present in small firms (ref. 7), and the subsequently high level of product innovation (ref. 8), is produced by the close link between design and construction as research and development engineers liaise with their constructing counterparts on the shopfloor. Indeed, this close-knit work environment is the arrangement that large firms seek to emulate through their various "intrapreneurship" schemes (ref. 9). Moreover, in practice, small high technology firms that are actively engaged in product innovation (as opposed to sub-contract work), rarely rely solely on external sources of product innovation, since such a reliance would mean that the important internal control of the innovation process is lost (see figure 1), and a minimum level of internal technical ability capacity must be maintained in order to interpret and subsequently modify externally acquired technical information.

The importance of some form of internal research and development capacity is bourne out by the observation that the high 85% level of internal research and development activity observed in the 1981-82 survey was broadly maintained in 1985-86 where the percentage of internal research and development facilities marginally declined to 79%. This overall picture of a marginal decline in aggregate levels of R&D activity in all three study regions will be reinforced by following data. Table 1 includes data for the incidence of internal R&D for the two surveys to indicate change over the five year survey period. Although the actual numbers of firms in the cells for the two surveys have reduced, the percentage changes in the three regions over the period have only marginally declined, with the largest decrease of 8.2% recorded in Scotland. It is notable that the numbers of firms

TABLE 1

Presence of R&D Activity in Survey Firms by Region 1981 to 1986

Presence of R&D	South-East England		Scotland		San Francisco Bay Area		Totals	
	81-82	85-86	81-82	85-86	81-82	85-86	81-82	85-86
	n %	n %	n %	n %	n %	n %	n %	n %
Presence of R&D	51 (85.0)	37 (80.4)	42 (77.2)	29 (69.0)	55 (91.7)	37 (86.0)	148 (85.1)	103 (78.6)
No R&D	9 (15.0)	9 (19.6)	12 (31.0)	13 (31.0)	5 (8.3)	6 (14.0)	26 (14.9)	28 (21.4)
Total	60(100.0)	46(100.0)	54(100.0)	42(100.0)	60(100.0)	43(100.0)	174(100.0)	131(100.0)

In 1981-82, n = 174, In 1985-86 N = 131

in the three regions with no R&D have not significantly declined during the five year period, which might suggest that the marginal decline in the percentages for regional firms with R&D are due to a disproportionate amount of non-participants in the 1981-2 category of firms with internal R&D. However, table 2, which indicates the breakdown of the closed and non respondent firms for the three regions reveals that the firms that did not participate in the 1985-6 survey were not over-prevalent in the 1981-2 category of firms with R&D. Indeed, table 1 reveals that the 85% - 15% respective balance between internal R&D - no R&D, was followed closely by subsequently closed firms as they answered for presence of R&D in 1981-82. This was 84% for those with R&D and 16% with no internal R&D facilities. The marginal reduction in firms with R&D must be jointly explained by both a loss of internal R&D in surviving firms, and the closure of firms.

TABLE 2
Level of Internal R&D Effort in 1981 for Closures and Non-participants in the 1986 Survey

Presence of R&D	South-East England		Scotland		San Francisco Bay Area		Total	
(N=43)	n	%	n	%	n	%	n	%
Internal R&D	12	(85.7)	10	(83.3)	14	(82.4)	36	(83.7)
No R&D	2	(14.3)	2	(16.7)	3	(17.6)	7	(16.3)
Total	14	(100.0)	12	(100.0)	17	(100.0)	43	(100.0)

A summary of the above data must generally conclude that, while internal R&D remains a strongly predominant means of generating technical knowledge in the small high technology firms of all the study regions, the presence of internal R&D facilities does not appear to offer survey firms any additional protection from closure.

Such an observation reinforces earlier comments which implied that the presence of R&D is poorly directed, the presence of R&D activities is a necessary but not sufficient requirement for successful innovation in high technology small firms. Indeed, if highly expensive R&D activities are poorly directed, the presence of R&D may hasten the demise of the ambitious small firm! This theme is leant further support by table 3 which compares the presence or absence of internal R&D with the generally valid performance variable which indicates percentage change in turnover (or gross sales) per employee over the five year study period. There is no support for the hypothesis that the presence of internal R&D produces a better performance in percentage change in turnover per employee. Indeed, the internal R&D category recorded an above average 87% of those firms recording a decrease in turnover per employee, while the no R&D category recorded a slightly above average 21% in the over 100% growth category. Not surprisingly, the resultant Chi square significance level for table 3 was an indifferent p=0.66%.

TABLE 3
Presence of R&D by Percentage Change in Turnover per Employee 1981-1986

Presence of R&D	Percentage change in turnover per employee 1981-86							
	Decrease		0-49		50-99		100+	
(N=117)	n	%	n	%	n	%	n	%
Presence of R&D	13	(86.7)	26	(74.3)	32	(84.2)	23	(79.3)
No R&D	2	(13.3)	9	(25.7)	6	(15.8)	6	(20.7)
Total	15	(100.0)	35	(100.0)	38	(100.0)	29	(100.0)

(Chi sq p = 0.66)

Table 4, which indicates employment growth over the study period, and not the improved efficiency implied in increase in turnover per head, reveals a slightly different picture from that of table 3. While those firms with no R&D are more prevelant than the average for the decrease category, and markedly less prevelant in the 50-99% employee growth category, firms with no R&D performed at above average levels in the over 100% employee growth category. Thus, the above trend of no clear pattern of inferior performance among firms with no R&D is continued.

TABLE 4
Percentage Change in Employment 1981-1986 by Presence or Absence of R&D

Type of R&D	Percentage change employment 1981-86								Total	
	Decrease		0-49		50-99		100+			
(N=131)	n	%	n	%	n	%	n	%	n	%
Presence of R&D	36	(76.6)	39	(81.3)	14	(93.3)	14	(66.7)	103	(78.6)
No R&D	11	(23.4)	9	(18.8)	1	(6.7)	7	(33.3)	28	(21.4)
Total	47	(100.0)	48	(100.0)	15	(100.0)	21	(100.0)	131	(100.0)

The impact of full- and part-time internal R&D commitment

The differing merits of part-time and full-time R&D have been discussed in the introductory part of this chapter (see figure 1). However, it is worth noting prior to the presentation of the following data that the dichotomy between full-time and part-time R&D may be at least as large in "sophistication of management" terms as the difference between internal R&D and no R&D, noted above. Clearly, there is a considerable gulf between an owner-manager performing two or three hours of R&D per week in the evenings, and a full-time R&D section employing ten research engineers. Moreover, if it is accepted that many of the fast growing firms with no R&D are successful because they do not require internal R&D support (e.g. a jobbing sub-contractor), firms with part-time R&D may possibly be viewed as product-based firms that "need" internal R&D as indicated by their product orientation and part-time R&D status, but are unable or unwilling to fund the full-time commitment that principles of efficient management would dictate. Set in this context, it might be hypothesised that stronger performance differences might be noted between survey firms performing full- and part-time R&D.

When viewing the changes in the balance between full-and part-time R&D over the study period, it is notable that there have been marked declines in the part-time category for both South East England and the Bay Area, causing increases in the percentages of participant firms with full-time R&D in these regions (table 5). In the Bay area, eight of the eleven firms

TABLE 5

Type of Research and Development Performed by Region 1981-1986

Level of R&D effort	South-East England				Scotland				San Francisco Bay Area			
1981(N=174)	81-82		85-86		81-82		85-86		81-82		85-86	
1986(N=131)	n	%	n	%	n	%	n	%	n	%	n	%
Full-time R&D	20	(39.2)	19	(51.4)	27	(64.3)	18	(62.1)	29	(53.7)	23	(62.2)
Part-time R&D	31	(60.8)	18	(48.6)	15	(35.7)	11	(37.9)	25	(46.3)	14	(37.8)
Total	51	(100.0)	37	(100.0)	42	(100.0)	29	(100.0)	54	(100.0)	37	(100.0)

lost from the part-time category were closed, while in the South East, the thirteen firms lost from the category included eight closures. Thus, the closure of firms with part-time R&D during the five year study period would appear to be a major determinant of the decline in part-time R&D in the South East and Bay Area regions. These data initially suggest that part-time R&D might not be an adequate level of innovation in competitive and innovative high technology sectors. Further evidence on the overall performance of firms with either full- or part-time R&D activity is provided by table 6 which relates these categories to changes in turnover per employee. Although the Chi square test on table 6 is insignificant at the p=0.05% level, it is clear that the group of firms with full-time R&D range from a below average 38% in the decrease category to 61% in the over 100% increase division, while those firms with part-time R&D comprise 61% of the firms in the decrease category and 39% in the over 100% grouping. While these differences are not sharp enough to produce a significant Chi square value, they generally suggest a better performance among those firms with full-time R&D.

TABLE 6

Type of R&D Performed by Percentage Change in Turnover per Employee 1981-1986

Level of R&D effort	Percentage change in turnover per employee 1981-86							
	Decrease		0-49		50-99		100+	
N=94)	n	%	n	%	n	%	n	%
Full-time R&D	5	(38.5)	13	(50.0)	20	(62.5)	14	(60.9)
Part-time R&D	8	(61.5)	13	(50.0)	12	(37.5)	9	(39.1)
Total	13	(100.0)	26	(100.0)	32	(100.0)	23	(100.0)

(Chi sq p=0.4)

This final sub-section on internal R&D commitment examines the levels of employment in those survey firms with a full-time R&D staff. The data presented in table 7 take the evidence for firms with full-time R&D staff beyond a crude measure of presence or absence to give precise employment size categories, and evidence of change between 1981-82 and 1985-86.

TABLE 7

Employment in Research and Development by Region 1981 to 1986

Number of R&D personnel	South-East England				Scotland				San Francisco Bay Area			
1981(N=76)	81-82		85-86		81-82		85-86		81-82		85-86	
1986(N=60)	n	%	n	%	n	%	n	%	n	%	n	%
1-4	16	(80.0)	11	(57.9)	21	(75.0)	10	(55.6)	15	(53.6)	15	(65.2)
5-9	4	(20.0)	5	(26.3)	4	(14.3)	4	(22.2)	6	(21.4)	4	(17.4)
10-19	0	(0.0)	3	(15.8)	3	(10.7)	2	(11.1)	4	(14.3)	2	(8.7)
20-49	0	(0.0)	0	(0.0)	0	(0.0)	2	(11.1)	3	(10.7)	2	(8.7)
Total	20	(100.0)	19	(100.0)	28	(100.0)	18	(100.0)	28	(100.0)	23	(100.0)

The main evidence of notable change occurs in the Scottish and Bay Area regions. Analysis of the closed firms for the study period indicates that Scotland lost seven firms in the under ten employee size band, while the Bay Area experienced the closure of six firms, with three of these enterprises in the over twenty R&D workers (employee) size category. Analysis of the efficiency of the firms in table 7 produced a very mixed picture in which there was no correlation of large scale R&D employment with large percentage increases in turnover per employee.

Management perceptions of the role of research and development

In the absence of detailed evidence, there is often a sweeping assumption by experts concerned with manufacturing innovation that the goal of technical information gathering in high technology small firms is the vigorous persuit of new product and process innovations in order to give the firm price, specification and production cost advantages in the marketplace. While intensive innovation may be the objective of such firms, this sub-section examines the regional incidence of this and other management strategies, and compares them with performance measures in an attempt to detect the efficacy of the various approaches to the generation of technical knowledge.

Table 8 reflects the spectrum of actual objectives displayed by survey firms in response to a question on the role of R&D in their firm. While in both the 1981-82 and 1985-86 surveys the very modest objective of merely "alleviating technical bottlenecks in production" was a rarely acknowledge approach, approximately one third of the respondents in both surveys claimed "incremental improvements in product and process design" as a rather modest main objective of their research effort. However, as might have been anticipated, approximately 60% of respondent firms in both survey periods mentioned "new product design" as their main objective. With regard to regional differences over time, the only notable change in the broadly similar pattern of regional responses in both surveys was in the South East of England, where 16 firms were lost from the 1981-82 "new product innovation" category. Further examination of this change revealed that seven of

TABLE 8

Level of Sophistication of R&D by Region 1981 to 1986

Sophistication of R&D	South-East England				Scotland				San Francisco Bay Area				Totals			
	81-82		85-86		81-82		85-86		81-82		85-86		81-82		85-86	
1981 (N=147) / 1986 (N=100)	n	%	n	%	n	%	n	%	n	%	n	%	n	%	n	%
Alleviation of bottle-necks	3	(6.0)	0	(0.0)	3	(7.1)	2	(7.1)	2	(3.6)	0	(0.0)	8	(5.4)	2	(2.0)
Incremental improvements	17	(34.0)	15	(41.7)	14	(33.3)	7	(25.0)	18	(32.7)	9	(25.0)	49	(33.3)	31	(31.0)
New product development	30	(60.0)	14	(38.9)	25	(59.5)	18	(64.3)	35	(63.6)	24	(66.7)	90	(61.2)	56	(56.0)
Other	0	(0.0)	7	(19.4)	0	(0.0)	1	(3.6)	0	(0.0)	3	(8.3)	0	(0.0)	11	(11.0)
Total	50	(100.0)	36	(100.0)	42	(100.0)	28	(100.0)	55	(100.1)*	36	(100.0)	147	(100.0)	100	(100.0)

* Final total does not sum to 100 due to rounding

these firms had closed, while 9 had moved to the other less radical categories of table 8. Although regionally specific, this evidence again points to the possibly risky nature of vigorous R&D towards new product innovation.

In keeping with the overall approach of this analysis, the management of the objectives of R&D are compared with the main performance variable (i.e. percentage change in turnover per employee) in table 9. Although a Chi square test on this table suffered from values of less than five in approximately half of the computed "expected" cell frequencies, the significance level was p=0.02% in excess of the normally accepted p=0.05% level. However, despite the problem of sparcely populated cells, the results are worthy of further analysis.

TABLE 9
Level of R&D Sophistication by Turnover per Employee 1981-1986

Sophistication of R&D (N=91)	Percentage change in turnover per employee 1981-86							
	Decrease		0-49		50-99		100+	
	n	%	n	%	n	%	n	%
Alleviation of bottle-necks	1	(9.1)	0	(0.0)	0	(0.0)	1	(4.3)
Incremental improvements	2	(18.2)	16	(61.5)	9	(29.0)	4	(17.4)
New product development	5	(45.5)	8	(30.8)	18	(58.1)	16	(69.6)
Other	3	(27.3)	2	(7.7)	4	(12.9)	2	(8.7)
Total	11	(100.2)	26	(100.0)	31	(100.0)	23	(100.0)

* final total does not sum to 100 due to rounding
(Chi sq p=0.02)

Since the respondents to the "alleviation" and "other" categories were negligible, a Chi square test was performed on the 78 firms in the "incremental" and "new product innovation" categories. This new test produced a Chi square value at the increased p=0.01% level of significance. The main cause of this increase in significance is the "bunching" of "incremental" firms in the modest 0-49% increase category, and the comparatively smooth increase in the proportion of firms in the "new product innovation" category, with a substantial number of firms in the over 50% and over increase category (table 9). This significant result is leant further importance by the observation that, although there is a smooth overall increase in the row proportions for the "new product innovation" category (i.e. 11% to 34%) from decrease to over 100% increase for the performance variable, the "new product innovation" motive also provided 45% (i.e. five firms) of the small number of cases in the decrease in turnover per employee category. Although the number of firms in the decrease category is small, the overall pattern of these results tends to suggest that the category of firms with the vigorous new product orientation for their R&D effort, contains both a majority of highly successful firms and a minority of very unsuccessful firms. If the introductory passages of this paper on the precarious nature of vigorous product oriented R&D are taken into account, this picture of contrasting fortunes is not particularly surprising. However, while this pattern of results is not self contradictory, it clearly undermines the assertion that R&D is always a beneficial activity.

Management and the control of R&D

The earlier work on which this follow-up study is based has stressed the cyclical nature of R&D expenditure and the financial stress points caused by periods of intensive R&D, often at times when revenue from sales is low (ref. 5). The introduction to this paper has again emphasised the risky nature of much R&D effort. There are a number of risks associated with the performance of R&D that are inherent to the activity and not conducive to elimination through some form of "optimal behaviour". Nevertheless, the following evidence on the manner in which small high technology firms control R&D suggests that part of the explanation for the failure of a number of firms that are heavily involved in R&D results from very poor management of the aims and cost-benefit of R&D activity. The first section of this analysis is concerned with a consideration of the fundamental management task of determining the objectives for internal R&D, which utilises data from the total survey, while later detailed evidence on the control of R&D funding is provided by more intensive personal interviews.

Table 10 indicates the manner in which the 100 survey firms responded to a question on the determination of R&D objectives. Given the dependent nature of the relationship of most small high technology firms to their frequently larger customers, it is perhaps not surprising that a majority of respondent firms stated that their customers were the instigators of internal R&D objectives. Much of the R&D work performed by survey firms directly results from the development of specific products for individual customers of which the custom built "one off" is an extreme example of a customer driven product innovation. Such a pattern of behaviour has both advantages and drawbacks. From a positive angle, the performance of R&D as part of a confirmed order for a product ensures that R&D costs can be written off by their inclusion in the quoted price. The price charged by high technology small firms is rarely an intensively competitive figure due to a lack of competition in the specialist niche production areas occupied by such firms. However, there are two major negative elements to this pattern of behaviour. First, it is clear that the actual final cost of R&D to the small high technology firm may substantially exceed the estimated cost included in the quoted price. And second, on a more general strategic note, an excessive reliance on the dictates of larger customers for the setting of internal product innovation goals may mean that responding to the specific needs of such firms results in the loss of opportunities to develop products that might have broader applicability and subsequent growth potential that exceeds the growth achievable by meeting the needs of such patrons. In this context, the results of table 10 prompt the observation that, since the firms operating internal control of the technical aims of the firm through boardroom decisions are in a <u>minority</u>, the strong reliance on customers for the setting of R&D objectives in all regions is evidence of a general lack of incisive R&D planning.

This analysis of the management of R&D now moves to a more detailed level and utilises the interview sub-sample of matched firms from the study regions. Since the financing of R&D has been consistantly asserted in the earlier paragraphs as a key area of expenditure for small high technology firms, respondents were asked whether their firms calculated the cost of R&D.

TABLE 10

Method of Determining R&D Objectives by Region 1986

Source of R&D objectives	South-East England		Scotland		San Fransisco Bay Area		Total	
1986 (N = 100)	n	%	n	%	n	%	n	%
Customer provides lead	22	(61.1)	13	(48.2)	19	(51.4)	54	(54.0)
Boardroom decision	13	(36.1)	13	(48.2)	13	(35.1)	39	(39.0)
Other	1	(2.8)	1	(3.7)	5	(13.5)	7	(7.0)
Total	36	(100.0)	27	(100.0)	37	(100.0)	100	(100.0)

Table 11 indicates that, although a majority of survey firms did calculate the cost of their R&D, a substantial minority (i.e. 38% of this subgroup of firms) made no attempt to monitor the cost of their R&D effort. There is a strong regional consistency about the average of 62% of firms that calculated the cost of R&D. However, while the calculation of R&D costs indicates a degree of control of expenditure on research within the firm, mere calculation does not necessarily imply that this exercise is part of any process of control, since such information may have been collected for other purposes (e.g. to claim tax relief).

Further light is shed on the incidence of R&D management and control of R&D expenditure by table 12 which indicates the proportion of personal interview respondents that attempted to evaluate the return on R&D spending. Only nine firms (19% of the total sub-sample) made an attempt to calculate the return on R&D expenditure. Moreover, seven of the nine firm respondents claiming to calculate R&D described the benefit to the company of sales which result from products which involve the type of R&D tailored to very specific projects, noted during discussion of table 10 above. However, in mitigation, it must be conceeded that the monitoring of <u>either</u> the costs or benefits of R&D expenditure is an extremely complex task in small high technology firms, where both human and physical resources may be used in more than one role to "cross-subsidise" several activities. Indeed, recent research on the calculation of the cost-benefit of research in very large firms did not produce any more impressive methods of calculation (ref. 10).

TABLE 11

The Incidence of Research and Development Cost Estimation by Region 1986

Estimation R&D costs	South-East England		Scotland		San Francisco Bay Area		Total	
1986 (N=45)	n	%	n	%	n	%	n	%
Estimates cost of R&D	9	(60.0)	9	(64.3)	10	(62.5)	28	(62.2)
Does not estimate cost of R&D	6	(40.0)	5	(35.7)	6	(37.5)	17	(37.8)
Total	15	(100.0)	14	(100.0)	16	(100.0)	45	(100.0)

None the less, both this evidence, and the general results from the previous survey of the study firms, suggests that the majority of these small high technology firms have little conception of the value they receive from the R&D effort they expend. Clearly, this lack of knowledge may be critical since such small firms may maintain large R&D commitments which, without evaluation of their contribution to the general well-being of the firm, may cause bankruptcy <u>before</u> their poor cost-benefit is detected. There is considerable evidence from the above analysis that extensive R&D is no insurance against a poor overall performance, or ultimate closure.

THE ACQUISITION OF EXTERNAL INFORMATION

The extent of external technical information contacts

The importance of local external technical information contacts to the innovation performance of high technology small firms has recently become an important issue for those concerned with high technology industrial growth. In particular, a major rationale for the burst of activity associated with the establishment of science parks attached to British universities has been the belief that small high technology firms will benefit

TABLE 12

Incidence of evaluation of return on R&D expenditure 1986

Evaluation of return on R&D	South-East England		Scotland		San Francisco Bay area		Total	
1986 (N=28)	n	%	n	%	n	%	n	%
Evaluates return	1	(11.1)	4	(44.4)	4	(40.0)	9	(32.1)
Does not evaluate return	8	(88.9)	5	(55.6)	6	(60.0)	19	(67.9)
Total	9	(100.0)	9	(100.0)	10	(100.0)	28	(100.0)

from technical links that will ensue from geographically close links with a university (e.g. see Cambridge University Science Park Directory (ref. 11)). In the absence of any strong proof that small high technology firms develop, or benefit from, strong links with local universities, this British enthusiasm seems to be based on the purely circumstantial American evidence of high technology small firm growth around MIT in Massechusetts and Stanford University in California (ref. 12). However, while such growth is undeniable, the causes of this small high technology firm birth and prosperity are unclear. Moreover, if there is evidence to refute the existance of strong technical links between American universities and small high technology firms, much of the rationale for university-linked British science parks is lost.

Recent evidence from the first phase of this research (ref. 5) has indicated that high technology small firms in both Britain and the United States rarely maintained important technical links with external information sources, while later work has questioned the causality of the "technical link theory" in explaining the growth of small high technology firms in the vicinity of the above mentioned American Universities (ref. 13). This work generally confirmed the results of earlier research (ref. 14), which indicated that the main thrust of technical information generation in such firms was channelled into the performance of internal R&D, and that external information sources were merely used to support internal R&D efforts (e.g. technical information on the performance of components and materials used in the construction of internal product innovations). This discovery should not be surprising given the well documented evidence that small high technology firms gain a significant innovation advantage over their larger counterparts through the close informal liaison of the R&D department with the adjacent shopfloor (refs. 6-8).

However, although the earlier stage of this study found little evidence of strong research links between small high technology firms and external information sources, the current phase of research offers the opportunity to note changes in the previous patterns, and compare the growth performance of firms externally active in terms of technical information gathering with those that rely solely on internal information generation. Table 13 combines results from the 1981-82 and 1985-6 surveys and indicates that there has been virtually no change in the overall proportion of firms with external information contacts between the two survey periods (i.e. a less than 1% increase). Table 14 indicates the 1985-6 survey firms' spread of contacts by type. While contacts with universities or other higher education establishments were clearly the most popular type of external involvement throughout the regions (i.e. 44%), industrial research establishments and other firms were important in the South East of Britain. Within the individual regions, there was a marked decrease from 56% to 36% in the originally impressive number of Scottish firms claiming external contacts. Of the 15 firm decrease, six of the losses were accounted for by closure. Conversely, the two other regions experienced small proportionate increases in the levels of their contacts with external technical information sources.

However, the earlier general impression that external technical contacts are not generally important is maintained in the 1985-86 survey by the 60% of firms with no external technical information contacts. The low general importance of external information contacts is confirmed by a further detailed question which poignantly asked firms with external technical information contacts if "serious disruption" would be caused to the firm's internal R&D effort by the loss of external information contacts. Only 19 (14%) of the survey firm total claimed that the breaking of their external technical information contacts would cause serious disruption to the firm's innovation performance. The

TABLE 13

The Incidence of Important Technical Information Contacts by Region 1986

Technical contacts	South-East England		Scotland		San Francisco Bay Area		Total	
1981 (N=173)	81-82	85-86	81-82	85-86	81-82	85-86	81-82	85-86
1986 (N=131)	n %	n %	n %	n %	n %	n %	n %	n %
Important contacts	25 (42.4)	23 (50.0)	30 (55.6)	15 (35.7)	14 (23.3)	15 (34.9)	69 (39.9)	53 (40.5)
No important contacts	34 (57.6)	23 (50.0)	24 (44.4)	27 (64.3)	46 (76.7)	28 (65.1)	104 (60.1)	78 (59.5)
Total	59(100.0)	46(100.0)	54(100.0)	42(100.0)	60(100.0)	43(100.0)	173(100.0)	131(100.0)

only significant regional change within this group of external information sensitive firms was an increase in South Eastern

TABLE 14

Type of Technical Information Source by Region for the 1986 Respondents with Technical Contacts

Type of contact	South-East England		Scotland		San Francisco Bay area		Total	
1986 (N=50)	n	%	n	%	n	%	n	%
University/ Higher Education	3	(13.6)	10	(66.7)	9	(69.2)	22	(44.0)
Private Consultant	1	(4.5)	0	(0.0)	0	(0.0)	1	(2.0)
Industrial Research Establishment	9	(40.9)	1	(6.7)	1	(7.7)	11	(22.0)
Government Research Establishment	2	(9.1)	2	(13.3)	1	(7.7)	5	(10.0)
Other Firm	6	(27.3)	1	(6.7)	1	(7.7)	8	(16.0)
Other	1	(4.5)	1	(6.7)	1	(7.7)	3	(6.0)
Total	22	(99.9)	15	(100.0)	13	(100.0)	50	(100.0)

* final total does not sum to 100 due to rounding

firms claiming that the loss of their technical information contacts would cause serious disruption to internal R&D, with an increase of 8 firms in this category to change their overall percentage of the South East regional total from 7% in 1981-82 to 26% in 1985-86. An examination of the performance of the two groups of internally and externally oriented firms over the study period did not produce any significant differences in their relative performances.

This analysis of external technical information contacts is concluded by consideration of the extent of the utilisation of licences and joint venture arrangements as specific means of acquiring technical information from external sources. On a superficial level, both forms of external collaboration <u>appear</u> to offer real opportunities for the small high technology firm to increase its resource base, and subsequent innovation performance, <u>without</u> the commitment of any substantial additional internal resources. However, there are specific problems associated with these strategies which require further discussion in conjunction with the presentation of evidence below.

With regard to the purchasing of a licence, the initial attractiveness of this approach derives from the ability to obtain a proven product without the need for risky R&D effort. In many industries, where the rate of technological change is comparitively modest (e.g. mechanical engineering), small firms may purchase the licence to a product (often from an overseas company) to good effect, ensuring a reasonable life cycle for the acquired innovation. However, in high technology industry, where product life cycles can be from three to five years, there is often little scope for the selling of licences, since by the time the original innovating firm has stabilised production of a potentially licensable product, this new innovation is in the process of being superseeded by more advanced technology. In such circumstances, the potential for licensing such a product rapidly declines. Moreover, since much of the technological content of high technology products lies in the constructional skills of highly qualified engineers, the products of high technology small firms are generally more difficult to transfer than their lower technology counterparts where the skills of construction are more universal.

Evidence from the 1985-86 survey confirms that licencing is not prevelant in any of the firms of the study regions. Table 15

TABLE 15

Extent of Technical Aquisition through Licencing Agreements 1986

Form of agreement	South-East England		Scotland		San Francisco Bay Area	
1986 (N=19)	n	%	n	%	n	%
Production link	5	(62.5)	3	(50.0)	1	(20.0)
Marketing agreement	2	(25.0)	2	(33.3)	0	(0.0)
Sales licence	0	(0.0)	1	(20.0)	1	(20.0)
Confidential	0	(0.0)	0	(0.0)	1	(20.0)
Other	1	(12.5)	0	(0.0)	2	(40.0)
Total	8	(100.0)	6	(100.0)	5	(100.0)

indicates that only 19 (14%) of the 131 survey firm total purchased licences in the five year period prior to the survey date. The main roles of these licences was either to permit the construction of another firm's products, or more simply, allow the survey firm to sell licenced product. The purchasing of a sales licence for a product offers two main advantages for the agent firm. First, problems over construction mentioned above are avoided by merely selling the new product. Second, the licenced product may be added to a firm's catalogue to fill gaps in its own product range and provide a commission at little cost.

The potential for the sharing of technical and financial resources is an initial attraction of joint ventures aimed at a number of potential projects which will assist the mutually advantageous growth of two firms. However, small high technology firms are less likely to enter into joint venture agreements than either lower technology or larger firms. Many of the above arguments of this chapter against the use of external technical information sources also apply to joint venture agreements. In particular, the critical advantage for small high technology firms of a close-knit team of R&D and construction engineers may be lost if a joint venture on product development is entered into by two firms. While the sharing of cost and risk burdens associated with expensive product development are attractive, the technical problems of performing R&D in two, often geographically distant locations, are formidable. Thus, for the most part, small high technology firms tend to work independently and internalise R&D effort in the knowledge that, although the cost of development work may not be shared, administration is far simpler, and any new innovation produced will be the <u>exclusive</u> property of the firm concerned. The universal unimportance of joint ventures is confirmed from survey evidence that indicates that only nine firms (7%) maintained a joint venture agreeement with another company. While five of these agreements concerned new product development, four of the remaining collaborations concerned the pooling of resources for marketing purposes.

These concluding investigations on licences and joint ventures lend further support to the general impression gained from this investigation of external information sources that the acquisition of external technical knowledge, when it occurs, is merely a secondary support to the most critical internal R&D effort. However, the wide ranging evidence of this paper has suggested that, although internal R&D is popular with the majority of survey firms, it is not necessarily a "passport to success". These issues will provide major themes for the following conclusions on the management of technical information acquisition in high technology small firms.

THE GENERATION OF TECHNICAL INFORMATION AND THE OVERALL PERFORMANCE OF STUDY FIRMS - A SUMMARY

Although the internal and external means of technical information generation presented in figure 1 are all valid potential means by which technical information may be acquired, the evidence of this second phase of the research into high technology small firms confirms the earlier work by generally emphasising the strong orientation towards internal R&D as the prime means of generating technical information. The lack of enthusiasm for external sources of technical information is reflected in the revelation that 60% of the current survey firms had no external information contacts; and that only 14% of those firms with external links considered them of sufficient importance that their loss would cause "serious disruption" to their internal R&D effort. The general lack of importance placed upon external sources of technical information was further supported by new evidence from the current survey on licencing and joint ventures which in both cases proved to be very rarely exploited options among the 1985-86 survey firms.

However, the application of the new time-series data to this latest phase of research has provided valuable novel insights into the relationship between the organisation and management of R&D, and the subsequent performance of survey firms. As the analysis progressed from the observation of the general to particular trends, a number of important results emerged. First, the possible difference in performance between firms with internal R&D and those with no R&D did not materialise. In discussion of this result, it was stressed that, for sub-contract firms with no requirement for internal R&D, problems over its full- part-time nature, objectives or control, were rendered truely irrelevant by a genuine and obvious lack of need, that allowed precious financial resources to be spent on other areas of the business to the overall good of the firm. This observation, together with an early acknowledgement that R&D was no automatic guarantee of success, helped account for the lack of a performance difference between these two groups of firms.

Nevertheless, a closer examination of the firms with internal R&D produced interesting evidence to indicate that, among those firms that required and undertook internal R&D, firms that adopted a more aggressive approach to such internal technical information generation tended to be more successful than those who adopted a more modest approach. In terms of both the commitment to full- or part-time R&D, and with regard to whether R&D was aimed at either incremental or new product development, firms with full-time R&D adopting a new product orientation with regard to their R&D motive proved to be generally more successful than their part-time, incremental counterparts, when related to the performance variables. There was some evidence to indicate that firms employing a part-time approach to R&D were less successful as indicated by their poorer performance in terms of growth in turnover per employee, and the abnormally high level of closure for this type of firm. With regard to such firms, it might be tentatively argued from these data that, while the existance of internal part-time R&D implies a need for internal innovation, part-time R&D does not represent an adequate level of commitment for a number of firms that must perform internal R&D. While an over-commitment to R&D may be folly, a reliance on part-time R&D in several high technology small firms in the 1985-86 survey may have proven insufficient to maintain the competitive edge of such firms in rapidly changing technologically complex markets.

Evidence from the intensive interview survey on the level of control of R&D exercised by survey firms confirmed, in hard data terms, a general impression that had been widespread during both the surveys on which this research is based. The R&D that evolves in high technology small firms is rarely a stongly planned activity. Initial survey evidence indicated that participant firms were strongly influenced in choosing the objectives for their R&D effort by the needs of their most important customers. While it would be foolish to underestimate the importance of larger customer need in determining the objectives of survey firms, in too many cases, the demands of customers merely fill a vacuum left by the failure of survey firms to consider formally what the role of their internal R&D

effort should be, including in particular, an unwillingness or inability to evaluate the opportunity-cost impact of various forms of R&D effort on the long term progress of the business. In this context, many high technology small firms of this survey are guilty of responding to market demand, rather than seeking to create new opportunities for the sale of new products. Problems caused by sub-sectoral recessions suggest the "short sightedness" of such an approach. In this sense, marketing and R&D are inextricably linked, since today's R&D project to develop a new product is a potential market success five years hence.

Detailed information on the control of R&D through the initial calculation of R&D expenditure, and the subsequent estimation of cost-benefit again revealed a surprisingly cavelier approach to R&D management. Considering the financial importance of R&D to the overall economic stablility of the firm, it is surprising to note that a substantial minority 39% of the sub-sample of interview firms with R&D made no attempt to calculate the cost of R&D, while only 24% of the sub-sample made any effort to evaluate the cost-benefit of their R&D effort. Clearly, an inability to monitor the benefit produced by cash invested in R&D may mean that short to medium term ineffectiveness of expensive R&D programmes may mean the long term demise of the firm. While R&D is an inherently risky activity, the above evidence suggests that many survey firms do not minimise the potential for poor results, since their approach to internal innovation is strategically and financially uncoordinated.

This paper has greatly benefitted from the availability to the current study of performance data with which various aggregate small high technology strategies can be compared. The results have exploded any myth that may have existed which argued that a strong internal R&D effort was <u>synonymous</u> with superior performance. The mixed nature of the results indicate that <u>both</u> R&D intensive survey firms and others with no R&D may be successful. The key to such success depends on the matching of technical information gathering provision as nearly to some "objective" need as possible, and subsequently ensuring that the level of activity is matched by its quality and appropriateness.

However, despite a substantial amount of poor management behaviour in the small high technology firms of this study, this paper has indicated that risk can never be completely removed from the R&D process. Since failed programmes of innovation in small firms may cause bankruptcy, there is considerable scope for government financial assistance with the "front-end" funding of such work. While the private sector may fund high technology R&D intensive small firms, the short term financial objectives of the venture capitalists and other private agencies diverge sharply from the wider needs of national and regional development. A stronger government commitment to the long term funding of <u>well planned</u> programmes of innovation in small high technology firms would allow for (hopefully rare) project failures, <u>without the collapse of the firms</u> which frequently transpires when venture capital investments are unsuccessful. Such funding could be linked to royalty payments on successful innovation projects to defray the cost of failed ventures. Thus, successful innovation sponsorship might offset failures in a government-operated non-profit making organisation which might exist within an existing aid agency (i.e. the Small Business Administration [SBA] in the USA and the Scottish Development Agency [SDA] in Britain).

The above arguments on support of R&D in high technology small firms are based upon the assumption that high technology small firms are not easily formed, particularly in depressed industrial regions. While there is considerable scope for the emergence of very badly managed firms with poor innovation and growth potential, firms with good prospects may <u>also</u> experience financial difficulties due to R&D costs exceeding original estimates, on projects that will eventually be very successful if the firm is allowed to continue in business. This problem is common to both large and small firms as evidenced by the "bailing out" of Rolls Royce by the British government in the early 1970's, when they experienced difficulties with the development of the, subsequently successful, RB 211 jet engine. If this strategy is warrented in the case of politically sensitive large firms, then the same principles apply to potentially successful small firms with a lower political profile.

However, this argument is not blandly based on the premise that money should be "thrown" at high technology small firms in a spenthrift manner. Indeed, there is considerable evidence from this paper that R&D is poorly managed in many high technology small firms. Any financial assistance towards R&D would clearly need to be strongly based on a clear programme of research that can be monitored and controlled by the firm's management and the loan agency. However, the loan agency should fully accept the basic truth that R&D is an unavoidably risky enterprise, and consequently, that failure is a potential outcome of the most proficiently run R&D programme. Critically if, after careful evaluation of the assisted firm and the failed programme of sponsored innovation, the agency considers that the firm concerned remains viable, the collapse of a single programme of innovation <u>should not</u> be allowed to cause the demise of the whole firm. Given this level of support, progressive high technology small firms with good growth potential would be more inclined to innovate, <u>and</u> would be more likely to survive the failure of individual R&D projects. In turn, this would be a major benefit to national and regional development agencies as they seek to increase <u>and preserve</u> the stock of high technology small firms.

REFERENCES

1. H.A. Simon, The role of expectation in an adaptive behaviouristic model, in: J. Bowman (Ed.), Experimentation, uncertainty and business behaviour, SSRC, New York, 1955.
2. H.A. Simon, Theories of decision making in economic and behavioural science, American Economic Review, 49 (1959) 253-283.
3. R.M. Cyert and J.G. March, A Behavioural Theory of the Firm, Prentice Hall, Englewood Cliffs, New Jersey, 1963.
4. M.J. Webber, Sub-optimal behaviour and the concept of maximum profit theory, Australian Geographical Studies, 7 (1968) 1-8.
5. R.P. Oakey, High Technology Small Firms, Frances Pinter, London, 1984.
6. R. Rothwell and W. Zegveld, Innovation and the Small and Medium-Sized Firm, Frances Pinter, London, 1982.
7. C. Freeman, Economics of Industrial Innovation, Frances Pinter, London, 1982.
8. J. Townsend, F. Henwood, G. Thomas, K. Pavitt and S. Wyatt, Innovations in Britain since 1945, Occasional Paper Series No. 16, Science Policy Research Unit, University of Sussex, Brighton, 1981.
9. E. Von Hippel, Successful and failed Co-operative Ventures: An Empirical Analysis, Industrial Marketing Management, 6 (1977) 163-74.
10. A.T. Thwaites and R.P. Oakey (Eds.), The Regional Economic Impact of Technological Change, Frances Pinter, London, 1985.
11. Trinity College Cambridge, Cambridge Science Park Directory, Trinity College, Cambridge, November 1983.
12. Technology Magazine, Industry Arrives at University, Technology Magazine, 8 (15) (1984) 26-27.
13. R.P. Oakey, British University Science Parks and High Technology Small Firms: A comment on the potential for sustained industrial growth, International Small Business Journal, 4(1) (1985) 58-67.

14. R.P. Oakey, High Technology Industry and Industrial Location, Gower, Franborough, Hampshire, 1981.

Innovation: Adaptation and Growth, edited by R. Rothwell and J. Bessant
Elsevier Science Publishers B.V., Amsterdam, 1987 — Printed in The Netherlands

SMALL FIRM LINKAGES IN THE UNITED KINGDOM

M N Beesley and R Rothwell
Design and Innovation Management Group
Science Policy Research Unit
University of Sussex
United Kingdom

SUMMARY

Collaborative activity between firms has become increasingly a focus for academics, policy makers and business strategists. Data from the SPRU innovation Data Bank indicate that small firms in the UK in particular have increasingly utilised external sources of technical expertise to aid their innovatory activities. This paper reports on a more detailed pilot study of the external inputs used by 103 innovatory small firms drawn from the innovation data bank. The data indicate very significant levels of inter-company linkages relating to firms innovatory activities.

INTRODUCTION

A recent phenomenon in some major sectors of industry (e.g. semiconductors, automobiles) is the rapidly growing number of international inter-company alliances (Hacklish, 1976; Sciberras, 1976). These "strategic partnerships" (SRI, 1986) take a number of forms including technological exchanges and joint ventures. They occur mainly between large, established corporations (Berg et al, 1982; Hladik, 1984). In addition, public policies in Europe, particularly those relating to information technology, are laying increasing emphasis on inter-firm collaboration, e.g. the Alvey Programme in the UK and the PAFE Programme in France (Rothwell 1985; 1986). At the same time EEC technology programmes mandate R & D collaboration between firms in different EEC member states, e.g. the ESPRIT Programme. Again, the linkages have occurred mainly between large firms. In short, to date, little attention has been given to studying inter-firm linkages in the small firm sector.

There are perhaps two main reasons why attention has been mainly paid to collaborative activity between large firms, as against the small firm: first, as Berg et al point out, detailed information on R & D activity is in general more readily available for larger firms: this is certainly the case in the United States where joint venture activity has been more widely studied. Such firms are, in the first place, more likely to gather R & D data systematically; second, organisationally, large firms are better able to engage in formal collaborative R & D activities, and such collaborations might therefore be expected to occur more frequently in the large firm sector.

Table 1 shows the advantages and disadvantages in innovation generally ascribed to small and large firms respectively. In terms of "plugging-in" to external sources of technology, small firms are seen to be disadvantaged vis a vis their large counterparts. Data on the sources of major technological innovation-initiating inputs to 2,200 important innovations introduced by British companies between 1945 and 1979 throw some light on this point. Table 2 shows, for small firms (employment less than 200) and large firms (employment greater than 10,000), the percentage of innovatory ideas deriving from internal and external sources for the periods 1945-1969 and 1970-1979.

It can be seen that during the earlier period small firms were indeed very much more introspective than their larger counterparts, just as predicted by conventional wisdom. During the latter period, however, the situation changed and small firms obtained a greater percentage of their innovatory inputs from external sources. Interestingly, while small firms obtained an average of 7% of their external ideas from 'other companies' during the period 1945-69, this had increased to 25% during the 1970-79 period.

While the SPRU innovation data indicated a growing level of inter-firm technological interdependency, it did not yield information on the relative sizes of the companies involved, i.e. the degree to which the technology transfers were from large to small firms or were between small firms, or on the factors underlying the apparently growing technological interrelationships. Nor did the data indicate any previous relationships between the donor and acceptor firms, i.e. supplier or customer relationships. The research described below, on inter-firm relationships in the UK small firm sector was designed to throw some light on these and related issues.

Table 1

Advantages and disadvantages* of small and large firms in innovation

	Small firms	Large firms
Marketing	Ability to react quickly to keep abreast of fast changing market requirements. (Market start-up abroad can be prohibitively costly.)	Comprehensive distribution and servicing facilities. High degree of market power with existing products.
Management	Lack of bureaucracy. Dynamic, entrepreneurial managers react quickly to take advantage of new opportunities and are willing to accept risk.	Professional managers able to control complex organisations and establish corporate strategies. (Can suffer an excess of bureaucracy. Often controlled by accountants who can be risk-averse. Managers can become mere 'administrators' who lack dynamism with respect to new long-term opportunities.)
Internal communication	Efficient and informal internal communication networks. Affords a fast response to internal problem solving; provides ability to reorganise rapidly to adapt to change in the external environment.	(Internal communications often cumbersome; this can lead to slow reaction to external threats and opportunities.)
Qualified technical manpower	(Often lack suitably qualified technical specialists. Often unable to support a formal R&D effort on an appreciable scale.)	Ability to attract highly skilled technical specialists. Can support the establishment of a large R&D laboratory.
External communications	(Often lack the time or resources to identify and use important external sources of scientific and technological expertise.)	Able to 'plug-in' to external sources of scientific and technological expertise. Can afford library and information services. Can subcontract R&D to specialist centres of expertise. Can buy crucial technical information and technology.
Finance	(Can experience great difficulty in attracting capital, especially risk capital. Innovation can represent a disproportionately large financial risk. Inability to spread risk over a portfolio of projects.)	Ability to borrow on capital market. Ability to spread risk over a portfolio of projects. Better able to fund diversification into new technologies and new markets.
Economics of scale and the systems approach	(In some areas scale economies form substantial entry barrier to small firms. Inability to offer integrated product lines or systems.)	Ability to gain scale economies in R&D, production and marketing. Ability to offer a range of complementary products. Ability to bid for large turnkey projects.
Growth	(Can experience difficulty in acquiring external capital necessary for rapid growth. Entrepreneurial managers sometimes unable to cope with increasingly complex organisations.)	Ability to finance expansion of production base. Ability to fund growth via diversification and acquisition.
Patents	(Can experience problems in coping with the patent system. Cannot afford time or costs involved in patent litigation.)	Ability to employ patent specialists. Can afford to litigate to defend patents against infringement.
Government regulations	(Often cannot cope with complex regulations. Unit costs of compliance for small firms often high.)	Ability to fund legal services to cope with complex regulatory requirements. Can spread regulatory costs. Able to fund R&D necessary for compliance.

* Note: The statements in brackets represent areas of potential disadvantage. Abstracted from Rothwell and Zegveld (1982)

SOURCE: R Rothwell, "Innovation and Firm Size: A Case for Dynamic Complementarity" Journal of General Management, Vol. 8, No. 3, Spring 1983

Table 2

Source of major initiating kmowledge inputs to 2,200 important innovations introduced in the UK between 1945 and 1979

	1945-1969		**1970-1979**	
	Firm Size		**Firm Size**	
	< 200	**> 10,000**	**< 200**	**> 10,000**
In-house	83%	58%	68%	77%
External	17%	43%	32%	23%

RESEARCH METHODOLOGY

Our research on inter-firm linkages has, to date, taken the form of a pilot postal questionnaire to all the small and medium sized firms in the SPRU innovation data base. This ensured that our initial survey was of a sample of innovative firms from all sectors of manufacturing in the UK. The results of the pilot postal questionaire are intended to form the basis of further, sectoral based surveys. From the 400 firms surveyed, we received 103 usable replies. In addition, 73 questionnaires were returned as a result of company relocations. This gave an effective response rate of just over 30%.

Initially the transfer of technology was perceived simply as being a relatively well defined and clear cut agreement to transfer a set of techniques, or knowledge inputs. However, on the basis of a series of pilot interviews conducted in ten firms taken from the innovation database, it was clear that there was a whole range of information inputs crucial to innovation which involve considerably more than a straightforward transfer of technology or a formal joint venture agreement. We isolated five areas of linkage which firms in our pilot interviews identified as being important to their innovatory activity. It was clear from the interviews that companies used activity in one area to gain access to expertise relating to another area of their activities; for example undertaking agency agreements in order to gain access to technology or to information on markets new to the firm. The areas of linkage identified were:

1) Contract-out R & D This is perhaps the most obvious source of technical

expertise for the small firm, either to gain access to technology and skills outside the firm, or simply to stretch scarce resources.

2) Joint R & D Ventures Formal or semi-formal arrangements with other firms jointly to develop a product or process of mutual benefit.

3) Marketing Relationships Agreements to act as agents for other firms' equipment in order to supplement their own product range, and open up new markets. This was, in some cases, a covert means of gaining access to new technologies.

4) Manufacturing Relationships These take a variety of forms: licencing of another firms' product can be a means of learning about a new area of technology; subcontracting manufacturing is perhaps the most standard way of gaining access to production processes and related skills not possessed by the firm.

5) Links with Educational Establishments, other public sector bodies and research associations These can take the form of sponsorship or placement of students during degree courses, and formal or informal access to a variety of facilities and technical expertise.

On the basis of the results of the pilot interviews, a postal questionnaire was constructed designed to elicit information on the range of inter-firm linkages listed above. In addition, since Government policy lays so much stress on assisting the small firm sector, we decided also to investigate the take-up by the sample firms of technology-related Government schemes.

The results of the survey relating to the area of inter-firm linkage identified as being important are presented below.

RESULTS

Overall, 89% of the firms had a significant link in at least one of the areas outlined above. Excluding links with public sector bodies and taking into account only those links with other firms, this figure drops to 84.5%. Looking only at technical and marketing links, 69% of the firms had some form of significant link. Finally, excluding marketing links, 47.5% of the firms had some form of technical link with other firms.

This clearly indicates that a large number of small and medium sized firms have some sort of external linkage which is of importance to their development of their business. A breakdown of the level of linkage by type is given in Figure 1. These linkage types are discussed separately below.

1) Subcontracted R & D

39% of the firms in the survey contracted out R & D. Of the 40 firms involved, 70% contracted out 10% or less of their total R & D; 14% contracted out between 11 and 20%; 2.5% contracted out between 21 and 30%; 85.7% of subcontracting R & D relationships were with firms employing less than 200; 8% were with firms employing 200 to 499; and only 5% were with firms employing more than 500. In terms of geographical distance, 18.4% of the subcontractees were located within 10 miles; 31.6% were located between 10 and 49 miles distance; 13.2% were located between 50 and 99 miles distance; 23.7% were over a hundred miles distance; and 5.3% were located abroad. 26% of the subcontractees were in the same sector as the contractor firm. In terms of vertical relationships 47% of the contractees were supplier firms, and 18.5% were customers.

In terms of the motives underlying the contracting out of R & D, in 34% of the cases, the main motive was to gain access to technology new to the firm, and in 50% of the cases the motive was to shorten lead times. 53% of the contracting out firms were forced to undertake such action because of a lack of in-house skills.

2) Collaborative R & D

26% of the firms engaged in some form of collaborative R & D venture. Most of the collaboration (56%) took place with firms employing less than 200. 46.6% of the collaborative firms collaborated with suppliers, and 34.6% with customers. 34.6% of the firms engaged in formal R & D Joint Ventures. However, a further 65% of the firms took part in collaborative R & D which involved active technical participation, not necessarily on a formal basis. 73% of the firms taking part in collaborative R & D played an active role in initiating the collaboration, with a similar percentage (74%) having a policy of actively seeking external ideas. The majority of firms with collaborative R & D (61.5%) already had established contacts with their R & D partners prior to the collaboration.

3) Marketing Relationships

I Agency

25% of all the firms in the survey acted as agents for other firms.

Figure 1

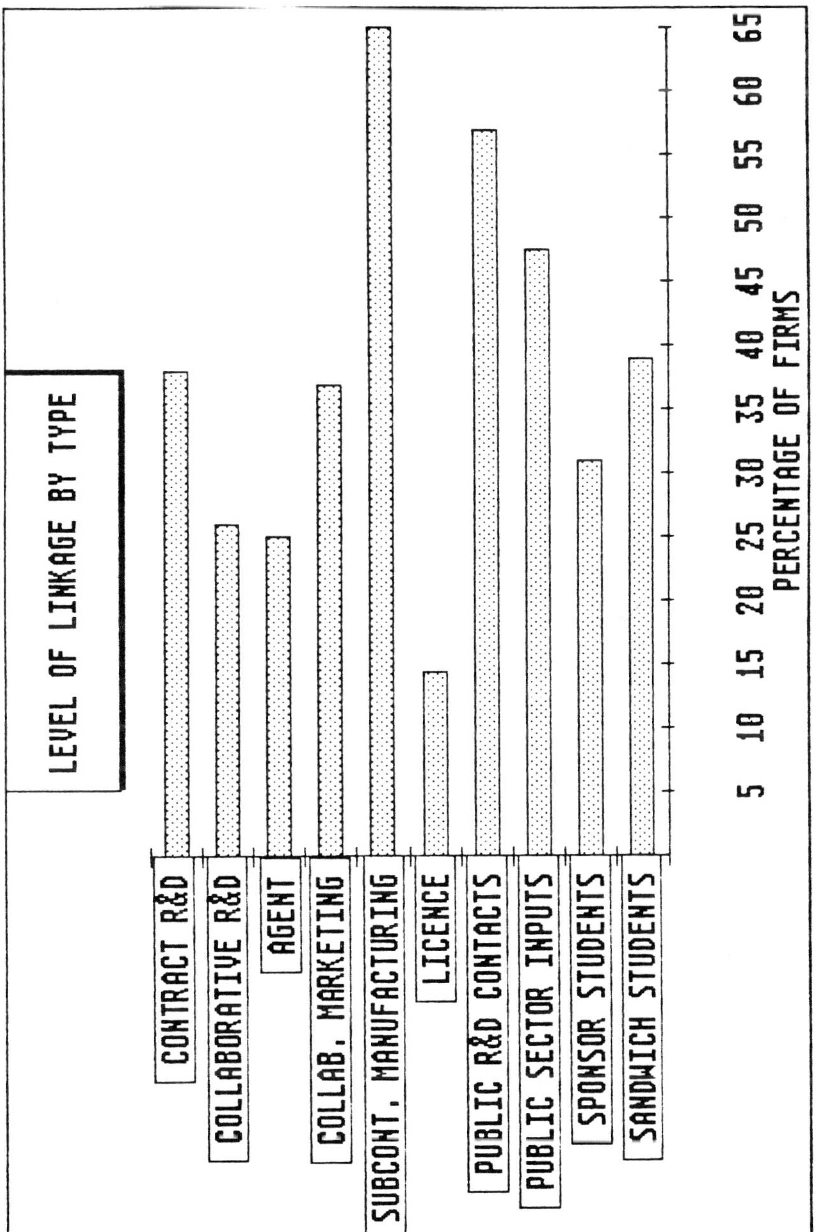
LEVEL OF LINKAGE BY TYPE
CONTRACT R&D
COLLABORATIVE R&D
AGENT
COLLAB. MARKETING
SUBCONT. MANUFACTURING
LICENCE
PUBLIC R&D CONTACTS
PUBLIC SECTOR INPUTS
SPONSOR STUDENTS
SANDWICH STUDENTS
5 10 15 20 25 30 35 40 45 50 55 60 65
PERCENTAGE OF FIRMS

Of this 25%, most of them had agency agreements with foreign firms (84%). The major reason for engaging in agency agreements was to gain access to products which complemented the firms' own product range (85%). This was effectively a way of making up for an in-house product development deficiency. However, 12% of the firms who acted as agents used the agency to gain access directly to new technological know-how, and 16% used them to gain knowledge of a new market area prior to innovating its own products.

II Collaborative Marketing

37% of all the firms were involved with collaborative marketing arrangements, the major proportion of which (72%) took the form of geographical marketing agreements. 46% of the firms with collaborative marketing had arrangements to sell mutually complementary products. A major reason for engaging in collaborative marketing was to gain access to new <u>sectors</u> of industry (74%), whilst 20% of the firms used the marketing agreements to gain knowledge of a new market area, and 26% to gain access to new technological know-how.

4) <u>Manufacturing Relationships</u>

I Subcontracting Manufacturing

For many small firms, subcontracting manufacturing can be an important means of gaining access to new production technologies, and can enable firms to innovate products requiring new production techniques, without having to invest initially heavily in expensive, sophisticated production equipment. A large proportion of all the firms (68%) subcontracted some of their manufacturing. 52% of these firms subcontracted less than 10% of their total production, but 20% subcontracted between 10 and 20% and 27% subcontracted over 20%. Most of the subcontracted firms were local, with 66% being situated within 40 miles of the contracting firm.

II Licencing

Only 16% of all the firms manufactured other firms' products under licence, accounting for under 20% of total production in 87% of the cases. In 60% of the cases, a major reason for licencing was to gain knowledge of new technological know-how; 34% of the firms were also using licencing as a means of gaining access to new market areas, and 27% to gain knowledge of new markets in order to innovate in-house products in the future.

5) <u>Public Sector Linkages</u>

55% of the firms had regular contact with research and development activities being carried out in public sector institutes, and 51% had technical inputs from the public sector of use to their innovation activities. 31% of all the firms sponsored students at educational establishments, and 39% provided placements for sandwich students. Sponsorship and placement often provide a useful means of gaining access to state of the art research and knowledge.

We also obtained data on the take-up of government schemes by the sample of firms. 40% of the firms took up at least one government funded scheme in order to gain access to external technical expertise. Of these firms, 57% took up only one scheme, 33% took up two schemes, and 10% took up three schemes. A total of 64 schemes was taken up.

6) External Linkages by Industrial Sector

We attempted to break down our sample into broad industrial categories: electrical and electronics; mechanical engineering; chemicals and food; and traditional industries. The overall pattern is for the "traditional" firms to enjoy fewer external contacts than firms in the other three categories (the data are not statistically significant because of the relatively small sample). It could be argued that companies operating in traditional areas have lower technological requirements than those operating in the other three industry groups.

7) Linkages and Internal Resources

There exists convincing evidence to suggest that a firm's propensity and ability to forge external technical linkages depends to a significant degree on its internal technical resources, most notably the in-house employment of technical specialists (Rothwell and Beesley, 1987). In order to test this contention we mapped out the volume of external technical contacts as a function of full-time R & D employment.

This analysis showed that, in general, the volume of external contacts increases with employment of R & D personnel. However, it is interesting that a relatively high percentage of firms employing no full-time R & D staff did successfully access Department of Trade and Industry schemes; the figure increases only slightly with increasing R & D employment. However, most of the DTI schemes taken up did involve the use of technical consultants to transfer well established technology and related techniques to firms which, even with relatively sparce in-house resources, they could readily assimilate. The same category of firms similarly effectively utilize the services of Research Associations whose R & D activity might, perhaps, be assumed to focus largely on the solution of industry-wide, process related problems and problems of standards, testing and validation, rather than on new product development which might require complementary in-house resources. The data generally support the contention regarding the importance of employing technical specialists in enabling SMFs to access external technical resources.

CONCLUSIONS

Whilst much more analysis of the data clearly needs to be done, there are a few conclusions we can draw from this initial analysis.

Whilst most attention currently is being focused on collaborative R & D in the large firm sector, it is clear from our pilot study that there is considerable innovation-related linkage activity amongst small and medium sized firms. As we have seen, a significant number of firms do engage in collaborative R & D ventures, in the same way as large firms. Others choose to conduct their external R & D links through a subcontracting relationship. However, our study has indicated that there are innovation-related links made through activities which, on the face of it, do not have any overt technical R & D content. Perhaps the most interesting area of activity in this respect is marketing links. Some of the firms we surveyed used marketing links as a means of gaining technical know-how by selling other firms' equipment. More important motives for developing marketing relationships with other firms were either to gain access to a new market, or to gain knowledge of a new market in order to facilitate entry later, with indigenously developed products. This raised an important point - and one which has often been made before in other contexts: innovation consists of much more than just solving the technical and physical problems associated with product design and development; it consists also of a whole range of activities prior to the actual act of physically producing the new product.

For a small firm, developing a new product can be a very risky activity. It demands the commitment of a large proportion of investment funds to a single project, together with the opportunity costs involved in committing scarce skills and time to one activity rather than another. Anything which can be done to lessen that risk is of great benefit to any size of firm, but is perhaps more crucial to small firms with fewer resources. Engaging in activities which, for a low cost, can increase the market and technical knowledge of the small firm, can form an important part of the whole innovation process. As we have seen, a significant minority of the firms in the pilot survey use external linkages in this way.

REFERENCES

1 Berg et al (1982) "Joint Venture Strategies and Corporate Innovation",

Oelgaschlager, Gunn and Hain, Publishers, Inc. Cambridge (Mass)
2 Hacklish C S (1986), "Technical Alliances in the Semiconductor Industry", New York Centre for Science and Technology Policy, New York University, February
3 Hladik K J (1985), "International Joint Ventures", Lexington Books, Lexington, Mass
4 Rothwell R (1986), "Innovation and the Smaller Firm", in W Brown and R Rothwell (eds), Entrepreneurship and Technology: World Policies and Experiences, London, Longman
5 Rothwell R and Zegveld W (1982), "Innovation and the Small and Medium Sized Firm", London, Frances Pinter
6 Rothwell R and Zegveld W (1985), "Reindustrialisation and Technology", London, Longman
7 Sciberras E (1986), "Scientific and Technological Exchanges and Applied Collective Research Among OECD Countries", Technical Change Centre, 114 Cromwell Road, London
8 SRI International (1986), "Strategic Partnerships: A New Corporate Response", Report No. 730

Innovation: Adaptation and Growth, edited by R. Rothwell and J. Bessant
Elsevier Science Publishers B.V., Amsterdam, 1987 — Printed in The Netherlands

INNOVATION STRATEGIES OF SMALL INDUSTRIAL COMPANIES

Jaap Docter and Casper Stokman

Research Institute for Small and Medium-sized Enterprise in the Netherlands,
Industrial Research Department, P.O. Box 7001, 2701 AA Zoetermeer, The Netherlands

SUMMARY

This paper is concerned with the relation between innovation strategies and the market on which operate small and medium-size companies. From the results of a survey among 53 Dutch manufacturing companies, some strategies are described, a distinction being made between consumer markets and intermediate-supply markets.

The company characteristics, market characteristics and the approach to the market are described, and the various steps of the innovation process examined. The innovation strategy followed appears to be closely associated with the market to which the companies are oriented and the type of product they manufacture.

The main conclusion is that there is no single optimum way to accomplish innovations; the wanted innovation of a product can be achieved by different innovation strategies.

INTRODUCTION

In the framework of the Technology and Economy project of the Ministries of Economic Affairs, Social Affairs and Employment, and Education and Sciences, the EIM[1] has undertaken to analyse the innovation process in Dutch medium-size and small enterprise. The inquiry comprises a large survey among small and medium-size manufacturing companies. The object of the present paper is to relate the innovation process to the market on which theaw companies operate. Most of the studies of innovation refer to countries outside the Netherlands and to large companies. Typically, their final purpose is to develop a (normative) innovation model, on the implicit assumption that there is an optimum method to accomplish innovations.

Cooper, recognising the shortcomings of that approach, has moreover pointed out that such most normative innovation models refer to product innovations in the area of consumption goods[2]. Indeed, far less advance has been made with normative models for manufactured products. Admittedly, some contributions have been made by technical experts, but they tend to examine the innovation process exclusively from a technological point of view, giving hardly any attention to the marketing of product innovations. Cooper suggests takinq several aspects into account in describinq an innovation process. For one thinq, the innovation process may fall apart into many different activities (steps), of which Cooper distinguishes no fewer than 20. For another, not all innovations require passing through all the steps. Lastly, different models can be developed for different innovation processes.

The conception of the innovation process as a tiered model underlies the method applied in the Dutch Industrial Innovation Project (PII). The primary objective of this project was to help small and medium-size companies in their innovative efforts, the secondary object to develop a methodology for innovation consulting. From the evaluation of the project, the tiered model can help to structure the innovation process, and many variations are possible[3]. The Industrial Innovation Project was concerned only with the search stages of the innovation process, companies being assumed capable of implementing the innovation on their own[4].

Going on from there, this paper describes the later stages of the innovation process, giving full attention to product development and market introduction. From the results of a 1986 survey carried out among 53 companies in six different sectors, a number of observed innovation strategies will be derived. The 53 companies were not a random sample, but chosen for the purpose: employing fewer than 100 persons, they satisfy the Dutch definition of medium-size and small enterprise; moreover they were known, from previous research, to have developed one or more product innovations.

'Strategy', as employed here, should be understood as the approach observed to have been followed rather than as one consciously chosen and well considered.
The definition of a product innovation is based on the Innovation Note (1979)[5], which describes an innovation as "the development and successful introduction of new or improved products". The notion "new" requires some qualification. A product may be new in the absolute sense of new to the world. It may also be a novelty in the Netherlands. Or it may just be novel to the sector. All three types are considered innovations[6].

A distinction between companies active on the consumer market and companies more oriented to the industrial market suggests itself. Typically, most intermediate suppliers tend to serve fewer customers, with whom they maintain a more direct contact, than producers of consumer goods. Intermediate suppliers may therefore be more product- than market-oriented. Different relations with customers may also entail different ways of product development and market introduction for the two groups of companies.

Innovation processes may not only vary with the different markets on which companies operate, but also with the nature of the product. Intermediate suppliers in particular deal with a large variety of products. Bekker and Havranek distinguish suppliers delivering products of their own product, and suppliers working mainly to the specifications of their customers[7]. Three types of company may thus be identified:

1. companies manufacturing consumption goods;
2. intermediate suppliers supplying articles of their own design;
3. intermediate suppliers working mainly to specifications.

With the help of the statistical technique of cluster analysis, a few likely innovation strategies will be worked out for each group.

Medium-size and small companies do not tend to conduct an explicit marketing strategy. As Bekker and Havranek point out, companies often lack a proper work plan, their main objective being to supply for the needs of a well chosen market[8]. An apparently haphazard approach to innovation is not necessarily an inefficient one; rather is it an approach that responds to the market which a company serves. Indeed, the innovation strategy followed is not an isolated element, but must be considered in the context of the market on which a company operates.

COMPANY CHARACTERISTICS

We will now briefly describe the sample of 53 companies examined in the Netherlands in terms of general company characteristics.

From table 1, 14 out of the 53 companies surveyed produce goods for the consumer market, 12 design themselves the products they deliver to other companies, while 27 companies supply mainly to specification. Table 1 also shows that most of the 14 companies manufacturing consumption goods belong to the carbohydrate, textile and chemical sectors.

TABLE 1

Company characteristics by type of market (percentage of companies)

	Consumer market	Intermediate supplies	
		own product	to specification
Number of companies	14	12	27
Sector:			
carbohydrates	14	17	0
textile	29	8	4
chemicals	36	8	4
plastics	7	8	15
appliances	7	42	30
electric equipment	7	17	47
Total	100	100	100
Company size:			
1 -50 employed	43	8	30
51-100 employed	57	92	70
Total	100	100	100
Age of company:			
25 years or less	21	42	56
over 25 years	79	58	44
Total	100	100	100

With the 12 intermediate suppliers with products of their own design, the emphasis is on the construction of appliances. The 27 intermediate suppliers delivering to specifications are found mainly in the sectors Appliances, Electrotechnics, and Plastic-Products.

Strikingly, intermediate suppliers tend to be larger than companies mainly occupied in the production of consumption goods. Besides, the majority of the latter group of companies date from more than 25 years ago, while suppliers working to specification, which are in majority less than 25 years old.

MARKET CHARACTERISTICS

The assumption that medium-size and small manufacturing companies tend to serve a small number of customers[9] is confirmed by the results of our investigation as far as companies delivering to specification are concerned. Indeed, these companies sell a large portion of their turnover (42 per cent) to only a few customers.

Moreover, the proportion of export in total turnover appears to be larger among intermediate suppliers than among companies manufacturing consumption goods. The results suggest that intermediate suppliers more often believe their market to be rising or fast growing than do manufacturers of consumption goods. A closer analysis of the 1986 turnover figures confirms that. In respect of 1985, producers of consumption goods realised an average rise by 14 per cent, intermediate suppliers with self-designed products 21 percent, and companies supplying to specification even 24 per cent.

MARKET APPROACH

To appreciate the influence of the market on the innovation process, it is important to know how companies keep in touch with customers and by what avenues they try to play the market. Table 2 illustrates the three points.

Evidently, companies oriented to consumer markets keep in touch with customers more through representatives than through the management or other employees. One third of these companies employ agencies (especially for foreign contacts). With intermediate suppliers, the role of the management is far more important.

With suppliers of self-designed products, managers and representatives appear to be of equal importance for keeping in touch with customers. Again, more than a third of the companies employ agencies.

Companies supplying to customers' specification rely heavily on their managers, who seem the most appropriate discussion partners on account of their (often technical) know-how. Half of the companies work with representatives; agencies are relatively unimportant.

Strikingly, marketing is done mostly by direct contacts with customers, fairs coming second. Companies supplying their own products tend to present themselves at fairs. The company's fame appears to be of special importance to intermediate suppliers, while producers of consumption goods often address their customers through advertisements and mailings.

TABLE 2

Market approach by type of market (percentage of companies)

	Consumer market	Intermediate supplies	
		own product	to specification
Number of companies	14	12	27
Contact with customers:			
representatives	79	75	48
management	43	67	81
other employeess	36	17	41
agencies	36	42	22
Marketing:			
direct contacts	71	67	70
stall on fairs	57	83	59
renown	14	25	41
advertisements	64	33	37
mailings	29	8	33
agencies	43	33	33
other	21	17	19

STEPS IN THE INNOVATION PROCESS

Before analysing how much time an entrepreneur spends on various stages of the innovation process, from the inception of a new product to its marketing, let us consider which steps are relevant to the three groups of companies.

From table 3, to many companies the market appears to be the source of inspiration. The general conclusion is that activities relating to the market are far more frequent among manufacturers of consumer products than among suppliers delivering to specification. Apart from deriving ideas from the market, that includes drawing up broad and detailed sales forecasts. Activities associated with the design of the product, such as technical invention and technical feasibility are particularly relevant to intermediate suppliers. Among suppliers to specification, about half the innovations achieved spring from technical inventions. Table 3 shows that licenses play only a limited role in the innovation process. Prototypes are constructed especially by intermediate suppliers. With regard to the testing of prototypes, no differences can be observed among the various groups of companies.

Trial production is most frequent with manufacturers of consumer products, and far less with companies supplying to specification. Adjustment of production means (mainly machines) is done in particular by companies supplying self-designed products.

TABLE 3

Percentage of companies passing through a certain stage, by type of market

	Consumer market	Intermediate supplies	
		own product	to specification
Number of companies	14	12	27
Idea from the market	100	75	85
Technical invention	29	25	48
Broad market exploration	73	58	41
Technical feasibility	43	58	67
Obtaining a licence	14	8	15
Product design	64	83	59
Sales forecast	36	8	7
Construction of prototype	79	75	78
Testing of prototype	79	75	74
Trial production	64	50	37
Adjustment of production means	29	50	22
Introduction	100	100	100

On average, the suppliers of consumption goods spend more time on marketing activities than companies supplying to specification (table 4). In absolute figures, however, designing the product takes the most time. After the decision to innovate, an average three and a half months are needed to make a good product design. The technical development of the product, that is, the construction and testing of the prototype, takes almost four months. The adjustment of the production process does not take much time. On an average, producers of consumption goods complete the innovation process in fifteen months.

Suppliers of self-designed products are the slowest innovators, needing an average of seventeen months. In the comments to table 3 we have already pointed out that with these firms, marketing is eclipsed by other activities. The time spent on the other activities is evenly distributed. After four months of designing, an average two and a half months are needed to make the prototype and three and a half months to test it. Four months are needed to adjust the production process. Indeed, with this group of companies, product innovation is not just a routine event. In that respect they differ from the others.

With companies working to specification, the innovation process takes, on an average, fifteen months. The accent is strongly on product development. The needs of the customers are clearly outlined, and the production process is geared to the constant developing of new products. The technical feasibility takes one and a half months to establish, and the design of the product as long. Two months are spent in constructing the prototype, three months on testing it. Finally, on average, one month goes by in the trial production.

TABLE 4

Average duration of the various steps in the innovation process, by type of market (in days)

	Consumer market	Intermediate supplies	
		own product	to specification
Number of companies	14	12	27
Idea from the market	23	15	3
Technical invention	1	0	15
Broad market exploration	75	77	4
Technical feasibility	15	61	45
Obtaining a licence	0	0	0
Product design	105	120	45
Sales forecast	15	0	0
Construction of prototype	20	72	75
Testing of prototype	90	105	92
Trial production	45	29	30
Adjustment of production means	8	120	0
Introduction	15	8	0
Average innovation period in days	445	517	440

INNOVATION STRATEGIES: A CLOSER INSPECTION

Consumer goods

14 companies in our sample are suppliers to the consumer market. More than suppliers to the intermediate market they emphasise market aspects (ideas from the market, broad market exploration and sales forecast). A closer analysis of the group of 14 companies brings quite a wide diversity in the innovation process to light. Cluster analysis reveals three identifiable strategies[10]:

a. prototype strategy;

b. product-design strategy;

c. market-orientation strategy.

Table 5 indicates the steps taken with the three strategies and the time spent on each, expressed as the ratio between the number of days observed to have been spent on a step and the number of days needed for the complete innovation. With that system of reckoning, the sum of certain percentages of time spent may differ from 100: the higher the sum, the more steps have been carried out simultaneously. From the table below, the first strategy (prototype) takes far longer to accomplish (15 months) than the two others, both taking 12 months.

TABLE 5

Relative duration of the innovation stages for manufacturers of consumption goods (in percentages of total innovation time)

Definition	Strategies		
	Prototype	Product design	Market orientation
Number of companies	5	4	3
Cluster value (R^2)	0.63	0.71	0.82
Idea from the market	14	0	10
Technical invention	0	5	1
Broad market exploration	9	6	14
Technical feasibility	2	1	14
Obtaining a licence	0	0	0
Product design	18	51	0
Sales forecast	0	4	5
Construction of prototype	35	2	5
Testing of prototype	43	26	5
Trial production	0	9	29
Adjustment of production means	0	0	11
Introduction	1	20	0
Average total innovation period in days	444	382	361

a. Prototype strategy. With this strategy, observed with five companies, manufacturing and testing of the prototype take up, respectively, 35 and 43 per cent of total innovation time. Before the construction of the prototype is taken in hand, far more attention is given to design than to market exploration. The average time spent on the product's marketing is negligible.

b. Product-design strategy. Four companies have adopted a strategy emphasising the designing of the product. So much effort seems to go into design that there is no need to make a prototype; testing can begin at once. Little time is spent on market orientation, so that the launching of the new product takes fully ten weeks.

c. Market-orientation strategy. Finally, the sample contains three companies which spend hardly any time on technical development or making and testing a prototype. Full attention goes to the the product's marketability. The idea derived from the market is allowed to mature before the market is explored. Quite some thought is devoted to technical feasibility, after which the trial production and adjustment of the production apparatus are immediately taken in hand.

Supply of self-designed products

The second group of companies in our sample supply to other companies more or less standardised products of their own design. Within this group of twelve, cluster analysis has identified two categories with slightly different strategies for developing ideas into products to be launched on the market. The two strategies can be typified as follows:

a. equilibrium strategy;

b. instant-production strategy.

Table 6 shows the stages passed through with either strategy and the time spent on each activity. Apparently, for both categories the total innovation period is longer than for suppliers to the consumer market: 16 months for the equilibrium strategy and just over 18 months for the instant-production strategy.

TABLE 6

Relative duration of the steps in the innovation process for intermediate suppliers of self-designed articles (in percentages of the total innovation period)

Definition	Strategies	
	Equilibrium	Instant production
Number of companies	7	13
Cluster value (R^2)	0.62	0.78
Idea from the market	4	5
Technical invention	1	0
Broad market exploration	13	0
Technical feasibility	12	22
Obtaining a licence	0	0
Product design	44	9
Sales forecast	0	0
Construction of prototype	13	22
Testing of prototype	24	14
Trial production	23	10
Adjustment of production means	13	70
Introduction	6	10
Average innovation period in days	484	640

a. Equilibrium strategy. With this strategy, adopted by seven companies, more or less equal attention is given to the main steps in the innovation process. Some nine weeks go by in the broad market exploration, which obviously does not warrant a veritable sales forecast. The longest time is spent on designing the product (about seven and a half months). Testing the prototype and trial production take about equal amounts of time, and proceeded in constant interaction with the market. With some companies of this cluster, the production equipment takes an average nine weeks to adjust before the product is definitely launched; the introduction is also given some attention.

b. Instant-production strategy. Three companies appear to be able to market a product without much attention for developments on the market. The strategy is directly oriented to the technical aspects of production: a technical feasibility inquiry, followed by the construction of a prototype and trial production. Quite a bit of attention is given to adjusting the production equipment before the product is finally marketed.

Intermediate supplies to specification

27 companies in our sample supply products made to clients' specifications. A closer analysis divides the group into three clusters pursuing different innovation strategies. The three strategies, all of a marked technical nature, can be typified as follows:

a. technical-feasibility strategy;
b. technical-equilibrium strategy;
c. product-design strategy.

Table 7 shows which steps are passed through in the three strategies and how much time is involved in each activity. The total innovation period appears to differ considerably among the three strategies. The technical-feasibility strategy takes about 15 months, the product-design strategy about 8 months and the technical-equilibrium strategy about 3 years.

a. Technical-feasibility strategy. Whatever the strategy, with suppliers of custom-specified articles the market orientation is negligible. Nor is there question of any real introduction into the market. The first strategy emphasises technical feasibility, and spend equal amounts of time on the prototype and the trial production. Probably, the client makes the design, leaving prototype and trial production to the intermediate supplier.

b. Technical-equilibrium strategy. The idea is developed, the product designed and the prototype made and tested in consultation with the client. With this strategy, the intermediate supplier takes care of the product design. The prototype is carefully tested, probably in co-operation with the client: the test takes longer than the construction. The product is really launched into the market.

c. Product-design strategy. With this strategy, evidently the client submits a technically feasible idea to the supplier, who concentrates on designing the product and making and testing the prototype. Unlike strategies a and b, this strategy spends more time on developing than on testing the prototype. There is no question of a trial production or adjustment of the production equipment.

TABLE 7

Relative duration of the steps in the innovation process for intermediate suppliers working mainly to specifications (in percentages of the total innovation period)

Definition	Strategies		
	Technical feasibility	Technical equilibrium	Product design
Number of companies	11	6	7
Cluster value (R^2)	0.67	0.71	0.85
Idea from the market	8	0	0
Technical invention	7	5	0
Broad market exploration	4	6	0
Technical feasibility	17	1	7
Obtaining a licence	0	0	0
Product design	5	51	39
Sales forecast	0	4	0
Construction of prototype	16	2	46
Testing of prototype	13	26	33
Trial production	14	9	0
Adjustment of production means	0	0	0
Introduction	4	20	0
Average innovation period in days	440	1063	236

CONCLUSIONS

In this paper we have investigated, on the one hand, the markets on which small manufacturing companies operate, and on the other, the innovation strategies they employ. The term 'strategy' does not mean that small entrepreneurs consciously plan and control the innovation process; often the procedure is implicit.

The principal conclusion is that there is not one single optimum way to realise innovations. Many different innovation strategies can lead to product innovation. For one thing, the innovation strategy adopted is closely associated with the type of market on which a company operates. The hypothesis that intermediate suppliers are more product than market oriented, is confirmed by the time spent on the various activities (steps) during the innovation process. In comparison with companies oriented to the consumer market, intermediate suppliers spend more time on technical activities.

Beside a distinction by type of market (intermediate supply market versus consumer market), a distinction by type of product has proved meaningful (final product versus supply to specification). Companies supplying goods to specification appear to maintain a more direct contact with their customers than companies making final products. That implies in concrete terms, that companies supplying to specification hardly need to explore the market before undertaking a product innovation. As soon as a certain customer has shown the need for a new product to satisfy his specific wishes, the development process in the company begins. The design stage takes hardly any time at all and the attention focuses on the technical development of the product.

Producers of self-designed articles, on the contrary, tend to find a previous exploration of the market necessary. Mostly the exploration is broad rather than detailed. Moreover, these companies spend considerably more time on product design than companies delivering to specification.

That innovation is by no means always a routine procedure is most evident with intermediate suppliers of self-designed products; on an average, they devote much attention to the adjustment of their own production equipment.

A final important conclusion is that even within the clusters identified by type of product and type of market, many different innovation strategies remain possible, different not only in length of innovation period, but also, and in particular, as regards the presence or absence of certain activities, and their duration.

NOTES

1 Economisch Instituut voor het Midden- en Kleinbedrijf.
2 R.G. Cooper, The new product process: an empirically-based classification scheme, R&D Management, vol. 13, no. 1, pp. 1-13, 1983.
3 J. Buijs, Innovatie en interventie, Kluwer, Deventer, 1984;
W.E. During, Innovatieproblematiek in kleine industriële bedrijven, Van Gorcum, Assen, 1986.
4 J. Buijs, Innoveren is te leren, M&O, 1986/4.
5 Innovatienota, Staatsuitgeverij, The Hague, 1979.
6 The 'novelty' criterion is also operated by A. Kleinknecht, Innovatiepatronen in de Nederlandse industrie, draft report, Maastricht, 1984, and J.A.A.M. Kok, G.J.D. Offerman and P.H. Pellenbarg, Innovatieve bedrijven in Nederland, Groningen, 1984.
7 J.N. Bekker and G.W. Havranek, Commercieel beleid op basis van planning, Metaal & Kunststof, pp. T6-11, April 1986.
8 J.N. Bekker and G.W. Havranek, op.cit., p. T10.
9 R. Brown, Marketing for the Small Firm, Holt, Rinehart & Winston, London, 1985.
10 Cluster analysis has been applied to the time involved in the various steps by means of the Ward method. This method aims at an equal distribution of the number of observations per cluster. On the basis of a minimum R^2-value of 0.60, 48 out of the 53 companies could be classified in clusters. The innovation process of the remaining five companies was too divergent for allocation to one of the clusters to be meaningful.

Innovation: Adaptation and Growth, edited by R. Rothwell and J. Bessant
Elsevier Science Publishers B.V., Amsterdam, 1987 — Printed in The Netherlands

INNOVATION IN MANUFACTURING MEDIUM-SIZE AND SMALL ENTERPRISE: KNOWLEDGE BREEDS PROSPECTS

Casper Stokman and Jaap Docter

Research Institute for Small and Medium-size Enterprise in the Netherlands, Industrial Research Department, P.O.Box 7001, 2701 AA Zoetermeer. The Netherlands

SUMMARY

This paper presents the results of an investigation into the diffusion of product and process innovations among 352 small and medium-size manufacturing companies in the Netherlands. The innovations, classified into stages of the diffusion process according to their degree of novelty, have been related to company, market, co-operation and information characteristics.

The main conclusion is that the diffusion of innovations is stimulated by adequate training, by co-operation and by the use of information networks. For small companies, information networks consist mainly of direct business contacts, for instance with customers and suppliers, completed by informal contacts. The role of government and research bodies and consulting agencies appears to be of minor significance.

INTRODUCTION

In response to the fast growing attention for technological developments and the position of the Netherlands in the international technology race, a Technology and Economy Programme Committee was installed in 1984 by the Ministries of Economic Affairs, Education and Sciences, and Social Affairs. Assuming that technological innovation is of crucial importance for a country's economic development, this Programme Committee set out to find answers to the following questions:

> What factors govern the relation between technology and economy, and what conditions must be created to achieve the most favourable effect of technological innovation ?[1]

The Committee has requested the Industrial Research Department of the EIM[2] to identify the factors which determine the innovation of the production structure of Dutch manufacturing industry, in particular among the smaller manufacturing companies. Several partial reports have already been issued[3].

The present paper reports on the results of a survey among 352 companies from six manufacturing sectors, undertaken to establish the diffusion of innovations in manufacturing medium-size and small companies[4]. The six sectors have been chosen by a large number of criteria, the most important being their position in the phase of the life cycle. In that way, a cross-section of manufacturing industry has been obtained.

The central question is:

Is there a relation between the external information acquisition by, and the diffusion of innovations among, medium-size and small companies?

To gain an insight into that relation, we have investigated how certain company, market and co-operation features are connected with the process of information and diffusion.

THEORETICAL ELABORATION

Rogers[5], a prominent specialist in diffusion research, defines diffusion as "the process by which an innovation is communicated through certain channels over time among the members of a social system". To most diffusion research, two questions are central:

1. what factors determine the acceptation (or introduction) of an innovation by an individual (or company) ?
2. what factors are decisive for the diffusion of an innovation ?

The former question refers mainly to the decision to introduce an innovation. In the terminology of our survey: what activities did the company undertake before introducing the innovation ? The second question is more relevant to the research object in hand, which is to know what information is used by innovating and non-innovating companies. Previous investigations set out to compute the adoption time of one single innovation, that is, the time elapsed between the moment of introducing one innovation into a certain population, and the moment of complete acceptation. After the computation of the adoption time, the sample population is generally divided into adoption categories typified as basic innovators, early adopters, early majority, late majority and laggards.

The present investigation follows an entirely different approach, in that it recognises three groups to be surveyed:

1. users of new products (consumers);
2. users of new production processes or machines (companies);
3. producers of new products (companies).

While in traditional diffusion research, questions about one specific innovation are posed to either of the first two groups, our method has been to interrogate a great many companies, representatives of groups 2 and 3, involved in many different innovations. Such an approach does not permit the computation of adoption times nor their use for classification purposes. To circumvent that problem, we have first established that
a (technical) 'innovation' is to be understood in the sense of the Economic Structural Note[6] and the Innovation Note[7], as "... the development and successful introduction of new or improved goods or production processes". To define what is to be understood by 'new' we have resorted to the operational concept of Kok et al.[8], who distinguished: new to the company, new to the Netherlands, and new to the world. Kleinknecht added yet another category, namely, new to the sector[9]. This division by degree of novelty (or innovation degree) enables us to define adoption categories according to stages in the diffusion process without needing recourse to the adoption time of the innovation. The first company to in-

troduce a world novelty leads the diffusion process; a non-innovating company is reduced to the last stage. The simultaneous examination of a large number of different innovations enables us to observe, at one single moment, the various stages of the diffusion process.

Diffusion has been shown in literature to be a complicated process, involving many explanatory factors[10]. While most other (empirical) investigations examine just one aspect, the present one takes up the problem in all its complexity, relating information about and diffusion of innovations to a large number of company, market and co-operation features. Figure 1 illustrates diffusion of innovations as a function of the availability of information networks, the characteristics of companies and markets, co-operating associations, and the interaction among those four factors. Subsequent sections will deal with the various factors and the way they determine, by their interaction, the diffusion of innovations.

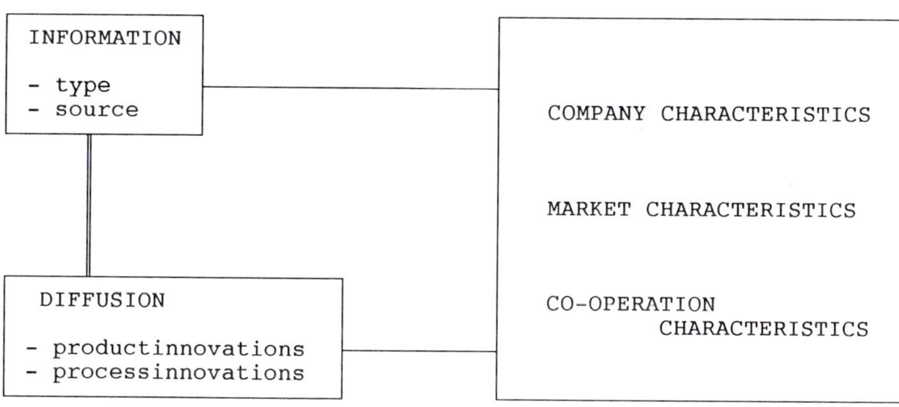

Figure 1. Relations in the diffusion process

DIFFUSION

Diffusion, in the terminology of this paper, is related to the innovation degree of the product or process innovation observed among the sample of 352 small and medium-size Dutch manufacturing companies employing fewer than 100 persons. Five categories can be distinguished:

companies with innovations

- new to the world
- new to the Netherlands
- new to the sector
- new to the company,

and

non-innovative companies.

The diffusion of innovations is supposed to depend on company, market and co-operative characteristics; companies at different stages of the diffusion process appear to show different combinations of characteristics. Highly innovative companies, that is to say, those whose innovations are new to the world, presumably show very special features.

We have reduced the number of categories of product innovators to four by classing companies with product innovations that are new only to the company with the non-innovative companies, on the ground that such innovations are merely product adjustments necessary for the company to survive. By those criteria, 63 per cent of the 352 manufacturing companies can be considered more or less innovative. Table 1 classifies the group of innovative companies by innovation degree and type of innovation.

TABLE 1

Innovative companies classified by type of and degree of innovation (in percentages)

Innovation type	Innovation degree				
	world	Netherlands	sector	company	(total)
Product	26	35	39	-	(100)
Process	3	11	24	62	(100)
All innovators	17	23	29	32	(100)

From this table, product innovations appear to be of a higher innovation degree than process innovations. No fewer than 61 per cent of the product innovations are new to the world or new to the Netherlands, which is not quite so surprising as it appears: most new products distinguish themselves sufficiently from existing ones to be unique in the world. Only 14 per cent of the process-innovating companies, however, have actually introduced innovations that are novel to the world or to the Netherlands. Apparently, the great majority of process innovations enters the company by diffusion. These figures are confirmed by another Dutch diffusion study[11].

The innovation degree, indicating how far the diffusion of innovations has progressed, is thus a criterion of the innovative capacity of either a given sector or a given country.

DIFFUSION AND COMPANY CHARACTERISTICS

Table 2 relates some company characteristics to the stage in the diffusion process.

TABLE 2

Companies classified by company characteristics and innovation degree (in percentages)

Company characteristics	Innovation degree world	Nether-lands	sector	company	non-innovative
Sector:					
modern	90	78	72	65	56
traditional	10	22	28	35	44
	100	100	100	100	100
Size class:					
2 - 19 employed	57	61	74	78	83
20 - 100 employed	43	39	26	22	17
	100	100	100	100	100
Education:					
employees with higher professional training	71	77	67	62	56
employees with scientific training	29	10	13	8	12
directors with higher professional training	45	43	38	27	16
directors with scientific training	6	5	7	2	4
Long-term projects	88	90	69	63	46

A distinction must be made between descriptive and explanatory company characteristics. The sector and size of the company are descriptive characteristics. Companies at the first stages of the diffusion process are larger and belong to more modern sectors than companies at the later stages, but that does not guarantee that a larger company from a modern sector is innovative. Other qualities, internal to the company, are needed. The main explanatory variables are the education and training of staff and management, the degree of planning, the management's attitude, the market approach, the ability to respond to co-operation openings, and the active use of information networks.

Table 2 illustrates the connection between the stage in the diffusion process and the training level of managers and employees. The stages 1 and 2 (new to the world and new to the Netherlands) contain more companies with higher-educated staff or managers than at the later stages. Some 44 per cent of these companies have directors with a higher education, the corresponding percentage among non-innovative companies being only 16. High-level knowledge appears to be a condition for the generation of important innovations and for leadership in the diffusion process. Probably, companies with high-skilled staff also use a different and more extensive information network than others. That point will be taken up later on.

Another fact emerging from table 2 is that at the first two stages of diffusion, the proportion of companies with long-term plans is twice that at the last stage. From that may be inferred that companies working according to plan and with a purposeful eye to the future, are better than others able to play an important part in the diffusion process. The same has been pointed out in other studies. Rothwell concluded from nine studies that companies with good planning and management techniques perform better than companies lacking these techniques[12]. From his report, management appears to play a crucial role in the diffusion process, in particular as far as combining internal possibilities with external information is concerned.

DIFFUSION AND MARKET CHARACTERISTICS

From the literature and from data from the Central Bureau of Statistics (CBS) we already know that relatively far more large than small companies are active in the export trade. We have investigated whether there is also a connection between exportation and diffusion of innovation. Table 3 indicates that the proportion of exporting companies is much larger among companies at the first stage of the diffusion process than among non-innovative companies. Complementary data show that not only the proportion of exporting companies, but also their export ratio is higher. The assumption that there is a connection between exportation to a world market and the generation of world-scale innovations is confirmed by table 3. Very probably, the higher educational structure and the intensive use of a specialised information network are among the causes.

TABLE 3

Companies classified by market characteristics and innovation degree (in percentages)

	Innovation degree				
Market characteristics	world	Netherlands	sector	company	non-innovative
Exportation:					
exporting companies	89	72	73	66	54
Market characteristics:					
sharp price competition	57	80	79	82	84
many product innovations	77	80	80	55	32
many process innovations	61	61	46	50	27
low profitability	56	69	66	61	69
high organisation degree	48	46	47	32	27

For a clearer understanding of the relation between innovation and exportation, a correction for size-class differences has been carried through. The conclusion from that exercise is that the proportion of exporting companies is clearly associated with the stage in the diffusion process, but that the share of exports in total sales is associated with the size of the company.

To establish a possible relation between market characteristics and diffusion, the companies have been asked to describe features of the markets on which they operate. Table 3 reproduces some of the results. From the analysis of the information emerges that the more innovative companies follow the developments in their environment, while non-innovative companies consider company-specific factors most typical of their market. Companies at the last stage of the diffusion process experience more competition in their market, are less profitable, create fewer product and process innovations in their sector, and are less adequately organised. They rely mostly on company-specific and financial factors. Probably, the lack of innovations with these companies is due to their inward orientation. Companies at the first stage appear to be less hampered by competition in their market. Their lead on other companies is large enough for them to be unconcerned about competitors. The creation of innovations and the organisation degree tend to decline as diffusion proceeds. These are important findings, which can probably be explained from the educational level and the external orientation of the companies. Subscription to sectoral associations or (international) organisations giving access to specific knowledge of the sector or technological developments applicable to the company, make for successful innovation and diffusion. To be accepted as a discussion partner, a company must possess expert knowledge, and that is what non-innovative companies lack.

The above conclusions shed new light on the significance of diffusion. Earlier investigations have mostly related the diffusion of innovations to company characteristics only. Our research has revealed the importance of market characteristics, in particular to innovators at the first stage of the diffusion process.

DIFFUSION AND CO-OPERATION CHARACTERISTICS

Current industrial and technological policy attaches more and more importance to the stimulation of joint ventures among companies. Co-operation is an explicit feature of the European technology programmes Esprit, Brite and Eureka. The technological policy pursued in the Netherlands also favours co-operation among companies and with scientific institutes; in the present 'interest-area' policy, co-operation ranks among the priorities. Because we do not yet know what activities are promising, and because promising activities are novel, co-operation across the sectors is essential. New forms of co-operation are needed to tackle the common problems of marketing and technology. Horizontal co-operation gives way to vertical and diagonal associations (for instance for the development and use of new materials). Business companies, government agencies and technical scientific institutes are partners in joint ventures.

The government much favours relations between large and small companies, preferably consolidated in a close association. The shorter life cycles of products, the increasing call for flexibility and shorter delivery times, require a reduction of the time spent on research and development. Co-operation can reduce both the investment costs and development time. The possibility of taking a lead in terms of market and production technology as well as know-how is an additional advantage; a possible monopoly gives a grip on customers and a vantage position on the market. As least as important are production or product spin offs: knowledge acquired in a co-operative association enables entrepreneurs to develop and launch new products.

In its inquiry, the EIM has found much difference in co-operative behaviour between innovating and non-innovating companies. An average 63 per cent of the former co-operate with other companies, against only 37 per cent of the latter, which points to a close connection between co-operation and innovation.

TABLE 4

Companies classified by co-operativeness and innovativeness (in percentages)

	Innovation degree				
Co-operation characteristics	world	Netherlands	sector	company	non-innovative
Co-operation with other companies	56	75	68	57	37
Areas of co-operation:					
market development	39	30	47	28	19
product development	30	42	26	40	14
process development	41	22	19	22	11
production under licence	5	18	11	9	4
purchase	13	12	5	4	7
other areas	14	3	2	7	4
Co-operation within the concern	83	77	66	45	43

With the exception of companies at the first innovation stage, co-operation appears to diminish as diffusion progresses. That companies with world innovations are not so eager to co-operate stands to reason: unique innovations do not lend themselves to co-operative efforts. Co-operation is most frequent with respect to marketing, products and processes. From table 4, first-stage companies tend to favour co-operation for the development of new production processes. Probably, these companies, with their high-level educational structure and extensive information network, are equals when it comes to developing new production techniques.

Among the first-stage companies belonging to a concern, 83 per cent appear to co-operate with others, against 43 per cent among last-stage companies. From complementary data we know that diffusion within a concern depends not only on the measure of co-operation, but also on the company's autonomy within the concern. Almost all last-stage companies that are part of a concern appear to have a specific task within the concern, and to have only limited freedom to decide about product and process innovation.

The conclusion is that co-operation enables companies to play an important part in diffusion. Without it, innovations might be slower to diffuse.

DIFFUSION AND INFORMATION CHARACTERISTICS

Information networks are important to diffusion. Most relevant Dutch studies have indicated the importance of information sources rather than their observed use. Nor have they linked the importance of sources directly to their specific uses[13]. By laying that link, our investigation gives an insight into the use of information networks on the one hand, and into the management areas for which specific information is used on the other. Table 5 surveys, for each stage of the diffusion process, the areas in which information is considered important and what sources are actually used.

In general terms, entrepreneurs tend to derive their product information from customers, suppliers, professional literature, and many other sources. Information on production processes comes mostly from suppliers, professional literature and fairs. Customers, professional literature, fairs, informal contacts and Chamber of Commerce are the suppliers of information about sales and distribution. Regarding information about competition, once more customers, professional literature and informal contacts, but also colleague entrepreneurs score high. Drawing conclusions about the various sources of information we should keep in mind that certain sources of information are relevant only to specific areas of management.

From table 5, product information appears to be considered the most, market information the least important. As diffusion progresses from stage one onward, information about aspects of management are found less important. The second half of the table shows that first-stage companies derive much of their information, apart from suppliers, from professional literature, fairs, TNO, universities and colleagues; to these companies, customers appear to be less relevant as a source. So, companies with new-to-the-world innovations favour external information sources, which are least popular among last-stage companies.

Consultants and universities score low as external sources of information, which is not surprising since our investigation addresses only the smaller companies (employing at most 100 persons). We know from other inquiries that the larger companies do indeed enlist the help of advisory bodies and consulting engineers; the smaller companies prefer suppliers as a source of information, for their advice is mostly free and a long-standing relationship has often fostered an (informal) bond of confidence, which is very important to small companies.

TABLE 5

Companies classified by information characteristics and innovation degree (in percentages)

Company characteristics	Innovation degree world	Nether- lands	sector	company	non- innovative
Aspects of management:					
product					
specification	69	61	64	64	40
quality	67	66	59	60	49
production process:					
process	60	59	39	58	37
raw materials/semi-products	76	63	70	62	64
market:					
sales/distribution	55	42	49	37	38
competition	44	44	54	53	41
Sources of information:					
direct business information:					
customers	39	52	44	57	48
suppliers	73	58	83	66	61
indirect business information:					
fairs	39	25	17	21	10
professional literature	60	45	41	53	21
research and advice:					
TNO/laboratories	10	8	2	12	6
universities	12	0	0	0	3
consulting engineers	5	8	7	2	4
inspection services	8	16	10	4	5
government/other:					
Chamber of Commerce	3	11	12	0	3
sectoral organisations	3	3	2	6	1
informal:					
informal contacts	26	18	24	40	15
colleagues	28	15	8	17	15

Complementary information has revealed which sources of information companies at various diffusion stages tend to use on which subjects. First-stage companies tend to ask their suppliers for information about the production process, either in direct contact or meeting them at fairs. Professional literature is important to these companies as a source of information on technical specification, quality of the product, and raw materials, but not on the production process. To latter-stage companies, suppliers are important too, but they also rely rather much on professional literature and fairs, and to a lesser degree on customers, informal contacts and colleagues.

Fairs appear to be important to first-stage companies with respect to their entire management; to other companies they are mainly relevant as sources of information on the production process. Besides, they offer opportunities for obtaining data on sales and competition.

Colleagues are among the most important sources of product information for first-stage companies; obviously, they can also give an insight into the activities of competitors. The other companies try to pump colleagues not only about the activities of competitors, but also about new processes. Informal contacts are especially useful to get information about the market. Strikingly, however, informal contacts are valued highest, for all information, by companies considering innovations which are new only to them. Indeed, the analysis has shown that latter-stage companies tend to go for free information from the more informal sources, including customers and sectoral organisations, while companies at earlier diffusion stages tend to consult more businesslike sources and seem readier to invest in information, enlisting the (paid) help of TNO, universities and inspection and consulting agencies. The use of these sources also depends on the distance - physical distance in particular - to be bridged. Sweeney's investigations, among others, have shown that distance is an obstacle especially to the smaller companies[14].

The conclusion may be that for medium-size and small manufacturing companies, professional literature, fairs, suppliers, customers and informal or formal contacts with colleagues are the principal sources of information and advice in the diffusion process. Engineering and advisory agencies and universities, while significant to the largest companies are for cost reasons and for lack of specific internal expertise hardly ever consulted by the smaller companies.

The assumption is warranted that the use which companies make of the external information network is largely determined by the educational level of directors and employees. Figure 2 illustrates first of all that highly educated entrepreneurs have wider information network at their disposal than lower educated ones. The latter primarily make use of direct businesslike sources, supplemented by indirect businesslike sources and informal contacts.

As the entrepreneur's educational level is higher, the information network is extended to such indirect businesslike sources as professional literature and fairs. Highly educated entrepreneurs tend to use more than lower educated ones research and advisory agencies. Their education enables them to exchange knowledge on a high level and accomplish the diffusion of innovations. On the other hand, they tend to rely less on informal contacts. The long-term planning they prefer calls for more formalised networks and reduces the need for informal contacts. Finally, whatever the entrepreneur's educational level, government and other agencies play a minor role.

Evidently, knowledge and information are important conditions for the diffusion of innovations and the ability to respond to prospects and needs of the market. To have the knowledge at hand is therefore important to a company. It may be stored in the form of documents, but may also be incorporated in the education, experience and understanding of staff and managers. The knowledge can be kept at par or extended by attending courses, studying professional literature, and visiting fairs and congresses. Remarkably, first-stage companies far more than latter-stage ones find their need for information inadequately provided for; they are also the ones which offer ample opportunities to staff and management for external orientation to innovations.

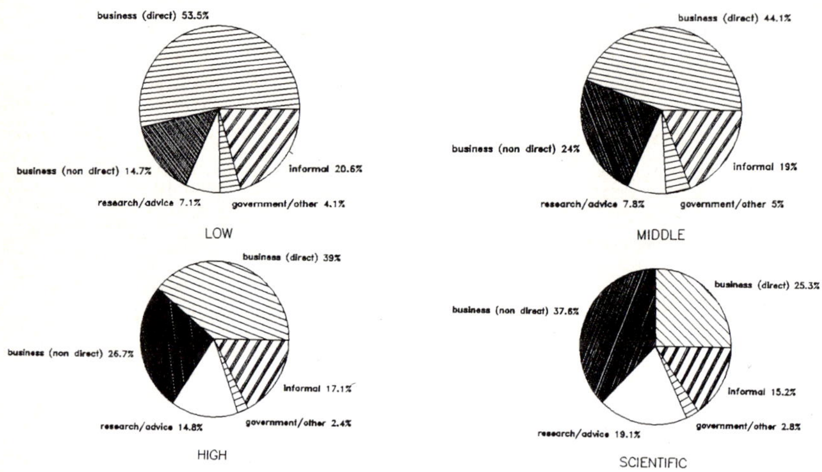

Figure 2 Educational level of directors and the information network

TABLE 6

Classification of companies by innovation degree and reschooling characteristics (in percentages)

Company characteristics	Innovation degree				
	world	Netherlands	sector	company	non-innovative
External orientation of employees:					
professional literature	95	73	84	69	55
fairs	95	63	76	71	57
courses	64	45	51	48	27
congresses/symposia	61	26	35	26	17
External orientation of management:					
courses/congresses/ symposia	86	61	59	52	41

From table 6, a very high proportion of companies at the first stage of diffusion seek external orientation. Last-stage companies are clearly least interested in orientation. The former group of companies are conscious that external information and good additional training are conditions for diffusion. The same follows from their need for additional information. First-stage companies look for new (future) information not only to suppliers, professional literature and fairs, but also, and preferably, to TNO, universities and advisory and engineering agencies. That the latter sources of knowledge are still little used is due to their poor accessibility, in terms of physical as well as psychological distance, to small companies and the high costs involved, particularly in consultation.

TO CONCLUDE

On the basis of the novelty of the product or process innovation we have been able to make the diffusion concept operational and to carry out analyses with respect to a large number of characteristics. From these analyses, the diffusion of innovation appears not to proceed in a uniform manner. There are numerous factors influencing, or associated with, the diffusion of new products as well as new production processes. A strong relation has been established between the stage of the diffusion process and the sector and size of the company. However, these factors only describe the companies at the various stages but do not explain the differences.

The most important factors to explain diffusion are the educational level of staff and management, active concern with and planning of future developments, the attitude of the management, co-operation with other companies whether or not within a concern, making use of additional training possibilities, giving opportunities for external orientation, and making an intensive and well considered use of internal and external information networks. Each of these factors can by itself contribute to a successful diffusion. The EIM study has revealed, however, that companies which combine them are best able to play an important role at the first stage(s) of diffusion. Thus, a 'business network' of many factors is created which can foster a company's innovative ability and its position in the diffusion process. Companies scoring only on one or a few of these factors are reduced to the later stages of diffusion, or even the very last stage, that of non-innovative companies.

These conclusions have policy consequences. For one thing, a good education is clearly essential. A higher education is the key to success and enables entrepreneurs to act as equal discussion partners with universities and research institutions.

Technical training courses should pay more attention to economic aspects. The very combination of technical and economic knowledge increases the chance of recognising innovation impulses. In medium-size and small enterprise, temporary associates may have a positive influence on the level of knowledge and the innovative capacity of the company. The fast technological progress renders additional schooling imperative. Company trainings and courses organised by sectoral organisations can make valuable contributions.
The French system of reducing company tax if a certain proportion of the turnover is spent on training and (re)schooling could give a powerful financial impulse.

The investigation has shown that co-operation can stimulate diffusion. Especially now that products tend to have shorter life cycles and the costs of development are rising, co-operation can reduce risks. Nor can a company keep an invention to itself for years and years: most probably, some competitor will soon enter the market with a similar invention.

Co-operation, albeit ad hoc, may offer a way out of such a predicament.

Co-operation with other companies and research institutions involves much transfer of knowledge. Government policy could stimulate such co-operation by subsidising contracted research. A suggestion would be to make the amount of subsidy dependent on the measure of co-operation.

So far, small and medium-size manufacturing companies hardly make use of universities, research institutions and engineering and advisory agencies as source of information with respect to diffusion. Research institutions are too inaccessible and engineering and consulting agencies too expensive to be enlisted as a source, which is regrettable because especially the latter group would be eminently suitable as an intermediary for the transfer of research knowledge to business companies.

When all is said and done, the entrepreneur himself remains the most important link in the diffusion process. A positive attitude towards diffusion, a high-level education, planning, regular retraining, and the capacity to make use of internal and external specific expertise, those are the factors which give a lead on other companies. The government can do no more than stimulating entrepreneurship in a general sense and for the short term by improving the entrepreneurial climate. However, technology fairs can serve to bring entrepreneurs in contact with new technologies and sources of knowledge and information, thus stimulating co-operation. In the long run, only education and training courses can offer a solution. Even then, the attitude and 'openness' of the entrepreneur determine whether new developments and possibilities of diffusion can be responded to.

NOTES

1 Technological and Economic Research Programme, Year 1984, June 22, 1984.
2 Economisch Instituut voor het Midden- en Kleinbedrijf.
3 A.H. Webbink, Groot en klein in de industrie, EIM, Zoetermeer, 1985; S.C. Adolf and C.T.M. Stokman, Vernieuwing in het industriële midden- en kleinbedrijf, University of Amsterdam, Department of Business Ecnomy, Amsterdam, 1985; H.J. Docter, Innovatie, informatie en diffusie, EIM, 1985; H.J. Docter and C.T.M. Stokman, Diffusie van innovaties: met kennis meer kans, EIM, Zoetermeer, 1987; H.J. Docter and C.T.M. Stokman, Het innovatieproces: een stapsgewijze benadering, EIM, Zoetermeer (to appear in 1987).
4 H.J. Docter and C.T.M. Stokman, Met kennis meer kans, EIM, Zoetermeer, 1987.
5 E.M. Rogers, Diffusion of Innovations, New York, 1983.
6 Economische Structuurnota, Staatsuitgeverij, The Hague, 1976.
7 Innovatienota, Staatsuitgeverij, The Hague, 1979.
8 J.A.A.M. Kok, G.D. Offerman and P.H. Pellenbarg, Innovatieve bedrijven in Nederland, Groningen, 1984.
9 A. Kleinknecht, Innovatiepatronen in de Nederlandse industrie, draft report, Maastricht, 1984.
10 H.J. Docter, Innovatie, informatie en diffusie, EIM, Zoetermeer, 1985.
11 Kok et al., op.cit.
12 R. Rothwell, The successfully innovative firm, European Conference on the Regional Environment for innovation and Technology, Barcelona, 25-26 February, 1987.
13 Information networks have been investigated by, among others: Kok et al., op.cit.; Nobin, Inventarisatie externe informatiebehoeften, The Hague, 1983.
14 G.P. Sweeney, Innovation, Entrepreneurs and Regional Development, Six Countries Programme, London, 1987.

Innovation: Adaptation and Growth, edited by R. Rothwell and J. Bessant
Elsevier Science Publishers B.V., Amsterdam, 1987 — Printed in The Netherlands

INNOVATION AND PRODUCTION: TWO TYPES OF FIRM BEHAVIOUR

P.R. BEIJE
Department of Industrial Economics, Erasmus Universiteit Rotterdam, P.O. Box 1738, 3000 DR Rotterdam (The Netherlands)

SUMMARY

To explain the different nature of a firm's production and innovation processes, a dichotomy between production-oriented and adaptive behaviour is introduced. Both behavioural concepts are based on assumptions about the nature of activities involved, the rationality of the decision-making process, and the environment in which activities take place and the information required for decisions has to be obtained. The main question is whether a firm's performances can be explained better by relating them either to adaptive or to production-oriented behavioural assumptions than to one integrated concept, like profit maximisation. Some examples of firm's performance are given to show the validity of the dichotomy.

INTRODUCTION

Economic literature on technological development reflects awareness of the trade-off between 'production' and 'innovation' on the level of firms (ref. 1). That is not to say, however, that a common type of analysis is used in this area; on the contrary: numerous approaches can be identified based on different definitions of production and innovation and different connections made between the two.

In the approach chosen here, production and innovation are defined in relation with the behaviour of different kinds of actor within the firm. In that respect it is worth mentioning that considerable attention has recently been given to assumptions on firms' behaviour which might explain different technological developments among and within sectors. Indeed, intra- and inter-sectoral differences in technological development cannot be 'explained' in what could be called 'production-dominated' analysis. Behavioural assumptions such as are made in the work of Nelson and Winter, Burton Klein, Pavitt (refs. 2-4), while naturally unable to take full account of the multitude of firm strategies and behaviour, do in our opinion at least deliberately avoid a 'one-exit' result (ref. 5).

The 'methodology' we have used to avoid a one-exit result is to make an 'ideal-type distinction' between two types of firm behaviour to explain different kinds of 'result'. The two types are called 'production-oriented behaviour' and 'adaptive behaviour'; production-oriented behaviour is applicable to allocation processes of given resources, adaptive behaviour is directed to the adaptation of existing and the creation of new resources.

With a view to that behavioural typology, further distinctions have been made between two kinds of activities, two kinds of rationality and two kinds of environment. With respect to the nature of activities we will look at the element of routine embodied in them, rationality is divided into substantive and procedural, following Simon's distinction (ref. 6), and an important aspect of the environment is how far actions taking place in it can be controlled by the firm.

Obviously, despite our analytical distinction, firm behaviour is mostly a mixture of production-oriented and adaptive behaviour. We will show that, when production-oriented behaviour is dominant, 'traditional' optimisation of known objectives is likely to be the relevant way to analyse performances. When adaptive behaviour seems relatively important, the analysis should shift towards the firm as a seeker of unknown opportunities, 'guided' by rather vague rules to plan and co-ordinate the behaviour of the employees involved. The main question can be formulated as follows: can different performances of firms be explained better by reference to either adaptive or production-oriented behaviour than by an analysis relating all performances to one and the same concept of behaviour ? And in addition: is it possible to formulate conditions which must be fulfilled to apply one or the other of the concepts ?

Starting from our behavioural typology, we will, among other things, define the learning curve as a reflection of 'learning-by-doing' within a controlled environment, and the creation of innovations as adaptive behaviour within a non-controlled environment; we will indicate possible intrasectoral differences as a result of the production-innovation dichotomy.

In the next section we will try to establish our position in the literature. Subsequent sections are devoted to the behavioural dichotomy and the eleboration of the concepts on the basis of their rational and informational aspects. Some consequences of the dichotomy for the analysis of different kinds of firm performance will conclude the article.

INNOVATION AND PRODUCTION

There is little doubt nowadays that a firm's profitability depends in the long run on its innovation performances or other adaptations to technological changes in its environment. On the other hand, profits from current production may, of course, largely determine the volume and nature of a firm's R&D-activities. We will use this section to indicate more precisely how our approach is connected with literature on the relations between innovation and production.

For one thing, our behavioural approach to the firm seems to root more in management science than in economics, as witness the work of Simon, Cyert & March (refs. 7, 8) and others. For another, our interpretation of 'environment' is important to understand the gradations of the dichotomy. While the concept

of a market structure concentrates on 'output relations' between incumbent and potential producers (ref. 1), to our mind, to analyse a firm's innovation processes properly, the concept of environment should be extended to the 'input side'; this 'input side', instead of as an exogenous set of technical production possibilities, should be seen as depending on the perception of the firm. In addition to the behaviour of the firm and the endogenousness of the environment, we will look at the nature of the different activities performed within the firm and especially to routines, elements which have also been used by Nelson and Winter (ref. 2).

Recently, Amendola and Gaffard made a distinction between the 'traditional' and 'evolutionary' approaches to innovation, which can be used to elaborate our line of thought (ref. 9). For that purpose we have constructed a table which shows the differences between, and some similarities of, the traditional and evolutionary approaches.

TABLE 1
Innovation and production in traditional and evolutionary approaches

	Traditional	Evolutionary
Innovation:	diffusion of innovations within a given economic structure	development of innovations within a 'technological paradigm' determined by given economic conditions
	- new given technology completely embodied in new given capital	- new and given basic knowledge 'applied' in in-house developed new capital equipment or products
	- given point of arrival of the process	- innovation and micro-economic structure are developed step by step; no a priori point of arrival
	- no technological environment	- a specific technological environment
Production:	transformation of generic inputs into finished goods	transformation of generic inputs into finished goods

Source: derived from the text in Amendola and Gaffard (ref. 9).

According to Amendola and Gaffard, the difference between the two approaches is gradual. Neither take into account two aspects of modern innovation processes which seem to be of increasing importance to firms: (1) the essential uncertainty of the environment, and (2) the role of specific labour inputs in the transformation process (ref. 9, p. 477). Both comments of Amendola and Gaffard will reflect in our distinction between 'production oriented' and 'adaptive'

behaviour.

OUTLINE OF A BEHAVIOURAL DICHOTOMY

A firm can in some respect be seen as a bundle of specialised activities (division of labour, technologically dictated skills) which have to be performed by individuals in a co-ordinated way. For our purpose it is not necessary to take detailed account of goals, strategies and organisational structures, although, as Porter has shown (ref. 10), a firm's performance depends to a considerable degree on the way activities are valued in relation to them. Individuals in the firm are supposed to behave economically by no other motives than those which made them decide to work for the firm in the first place. So, we define an individual's production-oriented behaviour as: the actions and/or co-ordinate activities which contribute to the production of a good or service and which are undertaken as part of a well-defined 'plan' in a known environment. Adaptive behaviour, on the contrary, is defined as the actions undertaken to search for, and research into, new technological opportunities; although the goals for 'adaptive actions' can be defined with relative clarity in some cases, it is assumed that the actions do not fit in a definable plan (the way that may lead to a new technical prospect is unknown in all cases), and that the environment is unknown and not controlled by the firm. On the level of firms, production-oriented behaviour consists of all production-oriented actions of employees that are, by defenition, part of the 'plan'. The plan itself presumably relates technical and organisational know-how, goals and strategies to one another to produce the current products in such a way that the production-oriented behaviour of each individual can be translated into a number of tasks, and procedures to operate them simultaneously and with predictable results.

While simultaneous actions of adaptive behaviour may also be operated within a framework of goals and strategies, the means to reach goals can not be 'translated' into precise tasks, and the results of simultaneously performed actions cannot be predicted in a way comparable to production-oriented behaviour. We will explain that presently.

Before we take up the characteristics of individual behaviour, one remark about the dichotomy so far presented may be useful: the difference between adaptive and production-oriented behaviour does not imply that an employee is necessarily occupied with either the former or the latter.

We regard production-oriented behaviour as 'routine', and adaptive behaviour as characterised by non-routine actions an decisions. On that score, we differ from Nelson and Winter, who distinguish important elements of routine behaviour in innovation processes (ref. 2).

The characteristics of production-oriented and adaptive behaviour are presented in the following table.

TABLE 2
Characteristics of production-oriented and adaptive behaviour

	production-oriented	adaptive
Nature of actions	routine fulfilment of a task in a larger whole	non-routine creative
Goal	production by given specifications	adaptation of current production
Co-ordination	part of routines	constant consultation and feedback
Result	predictable and controllable	uncertain

We will now return to the uncertainty of the environment and the role of human capital.

The environment where production-oriented behaviour is effected will be called 'production environment'. It is represented by buildings, capital equipment, etc., and could be extended to the operating and control procedures earlier called 'the plan'; the role of human capital is limited to specific tasks, based on experience and technical instructions. We assume that the co-ordination of individual activities is also a matter of routine. Uncertainty of the environment has been largely reduced to uncertainty about the results of the simultaneously performed actions according to the production plan; although some risk is always involved, we assume that the firm has acquired in the past a 'satisfactory' level of ability and experience to design a plan, a blueprint, for co-ordinated action, and to control the results of individual and simultaneous actions. It is in that context of known and experienced technical and organisational feasibilities that we speak of controllable and predictable production-oriented behaviour. A firm's performance depends, therefore, on its perceptions of market demand (how much to produce) rather than on how to produce. We will take up this point later.

The environment of adaptive behaviour can be called 'technological environment', a term which indicates that our present concern is only with the adaptation of production on the basis of technological prospects that, to the firm at any rate, are new. The uncertainty of the technological environment compared to the production environment can be explained from the lack of controllability

and the role of human capital. In production-oriented behaviour, the environment can change only as a result of actions and decisions of individuals participating in production, the role of the human capital being to perform tasks. In adaptive behaviour, on the contrary, the role of human capital is to perceive, and it is perception that largely determines the results of adaptive actions. The results will depend also on the actions of others outside the firm's control. Those are the elements which make us assume that the results of adaptive behaviour are unknown.

In a last step, we will extend the analysis of differences in the dichotomy to the rationallity of decision making and the role of information.

INFORMATION AND RATIONALITY

Every economic decision requires information, and economic actions are based, implicitly or explicitly, on decisions. Two aspects of information should be distinguished to serve our analysis: (1) the information on which a decision is based, and (2) the process of acquiring that information. Both can be linked to adaptive and production-oriented behaviour in specific ways, the result being almost invariably a kind of rationality.

To define the differences in information and rationality between production-oriented and adaptive behaviour, we will use the description of a decision process given by Arrow (ref. 11, p. 44). Arrow formulated the following properties of an economic decision process: "1. an objective function which indicates the relative desirability of different possible outcomes; 2. a range of policy alternatives, or instruments, among which the decision maker ... must choose; 3. the model which specifies the empirical relations connecting the instruments, the variables entering into the objective function and other relevant variables; 4. the computational methods by which the decision maker chooses the values of instruments so as to maximize the objective function subject to the conditions implied by the model".

It follows from our earlier description that adaptive behaviour is dictated completely by the plan which prescribes a mass of related decisions and actions. The information for all decisions is 'stored' in the plan, which could be seen as 'non-human and non-capital embodied knowledge' (ref. 12). Furthermore, co-ordination procedures provide a decision maker with information in the right place and at the right time. Middle management has often been said mainly to taken decisions based on information from different places in the firm, such as decisions concerning inventory schedules; such information may be called 'logistic'

Now let us consider adaptive decisions in the light of Arrow's properties. 1. The objective function does not include different possible outcomes but the

contribution of one outcome (or some related outcomes) to a larger whole; 2. as far as alternatives are relevant, the plan indicates which one to choose in which circumstances; 3. the model is part of the knowledge incorporated in the plan; 4. the plan also dictates a specific computational method to plan or control an action. The conclusion can only be that production-oriented decisions are a subset of Arrow's general set, with more limited properties.

It seems crucial that the production environment of production-oriented behaviour allows for the conception and use of a model in which all information for a sequence of decisions is available; Arrow calls it an 'information sequential' (ref. 12,p. 44).

In our opinion, the above description of the decision model of production-oriented behaviour is consistent with the rationality of economic decision making which Simon called 'substantive rationality': "Behaviour is substantively rational when it is appropriate to the achievement of given goals within the limits imposed by given conditions and constraints" (ref. 6, p. 130).

We will now analyse adaptive decisions in the same way. By definition, employees operating in the production environment have no access to the information required for such decisions; others have to obtain that information in the technological environment. We have assumed this environment to be the 'result' of perception by 'adaptive' decision makers, which makes the acquisition of information also a function of perception (see ref. 13 for a philosophical exposition). Or, in terms of the four properties referred to above: (1) there may be an objective function indicating the relative desirability of wanted outcomes, but mostly the outcomes of adaptive behaviour will not be known before actions are performed and completed; (2) in the same way, alternatives (instruments) are not fully specified; (3) in general there will be no quantative (empirically founded) model available; (4) a specific computational method cannot be chosen.

Again, the matter hinges upon the model: that the technological environment is uncertain means that 'the structure of the model' for decision making is unknown; in that case, according to Arrow, "... there is a cumulative growth of knowledge; as the numer of observations gets larger, there is an increase in the efficiency with which we estimate our parameters. "The information used in that way may be termed cumulative information (ref. 11, p. 48). In our case not only does the acquisition of cumulative information take time, but frequently, once a model of the technological environment (allowing, for instance, a 'substantive' decision process for technical innovations) is complete, new knowledge available to the 'perceivers' or important changes from outside make it useless. That the technological environment is not only 'unknown' but also uncontrollable implies that the process of 'cumulative information' depends on

individuals and organisations outside the firm and outside its control, such as other firms, universities, government organisations.

We already touched upon this point when we spoke about perception of the technological environment; perception in our terms calls for an effort by the people involved, it depends on knowledge acquired by experience and education, and it may change by research into new technological information - decisions of top management often rely on external information; in general no markets are available for that kind of 'strategic' information.

If we accept that decisions of adaptive behaviour are based on information from a constructed technological environment, and that such decisions may also change that construction, procedural rationality is obviously relevant, in particular in the light of Simon's definition: "Behaviour is procedurally rational when it is the outcome of appropriate deliberation. Its procedural rationality depends on the process that generated it" (ref. 6, p. 131).

To summarise: production-oriented behaviour is substantively rational and information for decision making springs from the controlled production environment; additional restrictions on perception of that environment make optimum results of actions following upon decisions possible; adaptive behaviour is procedurally rational, information being acquired by interaction among individuals in the firm as well as interaction with individuals in the technological environment outside the firm.

SOME CONSEQUENCES

We can only briefly discuss some important consequences as we see them, concerning the relations between behaviour and performance. Let us assume that following performance aspects of firms on the sectoral level are relevant to technological development: (1) differences in efficiency of production; (2) differences in the rate and/or economic significance of technical innovations; (3) a combination of the two. What, then, are the consequences of relating to them either the 'adaptive' or the 'production-oriented' concept ?

Differences in efficiency may in general be due to different production technologies or different performances with one and the same equipment. If two firms use different production technologies, again two explanations seem possible: in-house process innovation and heterogeneity of supply; we count the former to the second type of performance mentioned above. That leaves heterogeneous supply and different performances with one technology. Can we trace the conditions of behaviour ?

The behavioural aspect of different performances with a given production technology is commonly referred to as 'the process of learning-by-doing'. The conditions are: a specified production environment, and learning from informa-

tion that comes available in the very process of production; the firm's drive towards optimality may be assumed, making use of the production-oriented behavioural concept plausible.

When two firms in a market use different technologies, the conditions of behaviour are concentrated in the perception of the supply market of capital equipment. If that market is stable, we may expect well developed relations between our firms and suppliers; the production environment can probably be extended to the supply markets, and production-oriented - optimising - behaviour is the relevant concept. Different technologies might then be explained by properties of market structure, as is done in 'traditional' analysis. However, if the supply market is not transparant and stable, its perception is not a routine matter, and the analysis should shift to adaptive behaviour to understand differences in performance.

Performances in the creation of innovations might be susceptible to both concepts in the same way. If the firm can define the technological environment on the basis of long experience in the field and no 'drastic' changes expected, the conditions for production-oriented behaviour might be fulfilled. Amendola and Gaffard (ref. 9) refer to 'technological trajectory' in that way. We believe that in most cases these conditions are not fulfilled: the technological environment is constantly changing, and a firm's innovation performance depends largely on the effective perception of that environment. To explain them, these performances should then be analysed with the adaptive concept . An important aspect is the acquisition of knowledge about the technological environment (earlier referred to as cumulative information), which depends on the quality of R&D-employees and other factors determining the transfer of information.

The last 'performance' to be introduced is the combination of efficiency and innovation, an interesting point which we can only touch upon. As we have argued, individuals in an organisation may be occupied in both adaptive and production-oriented behaviour. The more production-oriented behaviour dominates, the more an employee's economic perception is likely to be limited to his or her section of the production environment, which reduces the chances of adaptive behaviour; a frequently heard opinion is that an organisation that concentrates on efficiency in the current production technology might lose its flexibility for adaptation in the long run. That takes us back to the raltion between production and innovation: adaptive behaviour can, in a dynamic setting, be seen as a trial (with errors to be expected) to create new 'structures', which may produce the conditions required for production-oriented behaviour with an acceptable level of efficiency and predictability from a managerial point of view.

REFERENCES

1. M.I. Kamien and N.L. Schwarz, Market Structure and Innovation, Cambridge University Press, Cambridge (Mass.), 1982.
2. R.R. Nelson and S.G. Winter, An Evolutionary Theory of Economic Change, The Belknap Press of Harvard University Press, Cambridge (Mass.), 1982.
3. B.H. Klein, Dynamic Economics, Harvard University Press, Cambridge, 1977.
4. K. Pavitt, Sectoral Patterns of Technical Change: Towards a Taxonomy and a Theory, Research Policy 13, 1984, pp. 343-373.
5. S. Latsis, Method and Appraisal in Economics, Cambridge University Press, Cambridge, 1976.
6. H.A. Simon, From Substantive to Procedural Rationality, in: S. Latsis (ed.), Method and Appraisal in Economics, Cambridge University Press, Cambridge, 1976.
7. H.A. Simon and J.G. March, Organizations, 16th. print, Wiley, New York, 1973.
8. R.M. Cyert and J.G. March, A Behavioural Theory of the Firm, Prentice Hall, Englewood Cliffs, 1963.
9. M. Amendola and J.L. Gaffard, Technology as an Environment: A Suggested Interpretation, Economie Appliquée, Tome XXXIX-1986 no. 3, pp 473-492.
10. M.E. Porter, Competitive Advantage, The Free Press (MacMillan), New York, 1985.
11. Collected Papers of Kenneth J. Arrow, vol. 4, The Economics of Information, The Belknap Press of Harvard University Press, Cambridge (Mass.), 1984.
12. F. Machlup, The Economics of Information and Human Capital, Princeton University Press, Princeton 1984.
13. E.E. Harris, Hypothesis and Perception, Allen and Unwin, London, 1970.

Innovation: Adaptation and Growth, edited by R. Rothwell and J. Bessant
Elsevier Science Publishers B.V., Amsterdam, 1987 — Printed in The Netherlands

AVAILABILITY OF TECHNICAL INFORMATION AS A SOURCE OF SECTORAL DIFFERENCES IN THE QUANTITY OF INNOVATIONS

P.R. BEIJE

Department of Industrial Economics, Erasmus Universiteit Rotterdam, P.O. Box 1738, 3000 DR Rotterdam (The Netherlands)

SUMMARY

Apparently, the acquisition and use of information by small and medium-sized firms is increasingly seen as an important aspect of their creation of innovations. The research has been restricted to technological information. Although the information process is probably highly complicated, a rather simple approach has been chosen. A positive effect of the presence and quality of an information structure on the number of innovations has been assumed; this structure comprises producers and transferrers of information. Firms in an industrial sector can be seen as competitors with a common information structure but different access and receptivity to it. A model for cross-section analysis is presented based on the assumption that different industrial sectors have different information structures which may account for different innovation performances.

INTRODUCTION

Currently, the creation of innovations as an important element of economic growth is seen as a major element of any industrial policy in European market economies; that 'revival' of innovations was reflected at the congress "Towards a competitive future for Europe" held in December 1985 (ref. 1). Apparently, the pace at which European firms are producing successful technical innovations is unsatisfactory in the eyes of many national governments; private firms in Japan, the U.S., and the N.I.C.s are producing at a higher rate, or at least the economic effects are more significant in terms of employment and productivity growth, etc.

The impression is that the disappointing lack of innovativeness of European industry is not due to a general low rate of scientific discoveries in Europe (ref. 2). If the 'production of new scientific knowledge' in Europe may be presumed to be on an acceptable level, then perhaps bottlenecks in the exchange of information between the 'scientific community' and the 'production system' must be blamed for the insufficient level of innovations.

We define 'scientific community' as a set of individuals and organisations which produce knowledge, and 'production system' as a set of organisations which produce goods or services; the production system uses scientific knowledge for part of the technological adaptations of production, and in doing so

creates technological knowledge.

One way to trace the importance of the dissemination of information from the scientific community (knowledge system) to the production system is to relate different rates of sectoral innovation to different qualities of that information dissemination. Admittedly, part of the knowledge system consists of firms belonging to the production system (for instance, R&D-departments of private firms), but the analytical distinction between production and knowledge systems does not seem irrelevant for small and medium-sized firms (smf), which often lack-R&D departments or are otherwise less than large firms occupied with science-based research.

This paper represents an attempt to construct a simple model for the estimation of cross-sectoral differences among 30 Dutch industrial branches of activity, on the basis of the quality of the exchange of technical information between the knowledge system and the sectors. But first we will treat some problems associated with knowledge as an economic good and the behaviour of firms with respect to external information.

KNOWLEDGE AS AN ECONOMIC GOOD

Once the economic importance of 'producing' technical innovations on the level of firms is accepted along with the hypothesis that the firm cannot create the technical knowledge required on its own, the economic importance of knowledge as an input to innovation cannot be denied. Things could only be otherwise if technical knowledge were a free good or a good already available within the firm (and applicable without opportunity costs !), in other words, if knowledge could be had for free and without time consumption. Obviously, firms do not, in general, possess all the information needed to innovate, and are therefore forced to acquire <u>external knowledge</u>.

How then, does a firm obtain the technical knowledge it requires ? If knowledge were a free good, competitors would have the same opportunity; blocking the use of some specific bit of knowledge by a patent might prevent that, but is a measure seldom resorted to. In a properly working knowledge market, the price of an identifiable piece of knowledge would reflect its production cost or useful value. To calculate the production cost, some notion of a production function is needed, which in most cases is lacking (ref. 3); one problem is to determine input costs, another the 'multiplicity' of knowledge: once produced it can be reproduced without additional cost (marginal costs being close to zero). The value of knowledge to its user is equally difficult to determine, because to know it one has to know the value of the output (that is, the value of the innovation itself and its market potential), which is hardly possible; moreover, the pre-estimated value, if it could be established, might well be

zero,since all potential innovators can acquire the knowledge (ref. 4). Such rather philosophical reflections follow immediately from the literature on information (refs. 3-4). Indeed, from the properties of information as an economic good ensues that an innovating firm's efforts to acquire knowledge cannot be treated as a process of market actions, unless we are talking about a licence or patent market.

We will treat knowledge acquisition as a process of active search by the firm as an integral part of the innovation process.

THE INFORMATION BEHAVIOUR OF A FIRM

Now that we are no longer concerned with the buying or selling of innovations but with a search of technical knowledge which might ultimately lead to an innovation, we encounter a well known paradox: one has to know before starting to search, but if one knows, search is no longer needed (our interpretation of a popular information paradox (ref. 5)).

To take up our analytical distinction between science system and production system: to initiate a search for knowledge, a firm must have at least a vague idea of the science system. Most firms probably lack a clear picture of the pieces of information they may find, although their ability to evaluate information may grow as specific knowledge within the organisation accumulates and individual experience with specific scientific fields increases. In that context, the difference between 'experiments' and 'terminal acts' in statistical decision theory is interesting to notice; experiments are concerned with collecting information for future terminal acts (decisions) (ref. 6). The purpose of this paper is not to analyse in detail the dynamic properties of a search process. To reflect some of the intricacies of the underlying process, we will understand by information any facts of which a firm cannot establish the probable economic value, and by knowledge all information of which the firm can state, with some probability, whether and how it contributes to the innovation process. As we said before, a firm looking for knowledge is likely to encounter all kinds of information in its environment.

With a view of comparing various sectors, we will look for 'structural components' in firms' search processes. We have made it clear that the search for information must be distinguished from the evaluation of information. An empirical investigation of innovations has produced the following factors governing the search for and processing of information:

- the quality and quantity of the in-house R&D;
- the intensity of employees' contact with sources in the environment;
- the exchange of information within a firm;
- the organisational structure and technical standard of information-proces-

sing equipment;

- environmental scanning as a part of a firm's strategy;
- the quality of technical employees (engineering);
- the availability of documentary information (professional journals, etc.).

These factors all seem to have been discovered by a detailed inquiry into strategic management and internal organisation. Since we are focusing on the external search process, we will not proceed in that direction, but abstract from organisational complexity by 'reducing' the above factors to two concepts of information: accessibility and receptivity.

A firm's receptivity is understood to reflect the technical knowledge and learning capacity it possesses, the organisation of its internal information processing, its propensity to change and its degree of internal interaction. The firm's access to information reflects its awareness of the technological environment, the organisation of its search process, the quality and intensity of its informal information exchange with the environment. Receptivity and accessibility together constitute the information profile of a firm, which determines what information will be found and how effectively it will be used (in terms of innovation). At this point a reminder is in order that even such abstractions from information behaviour does not relieve the scientist of serious data problems.

THE STRUCTURE OF INFORMATION SUPPLY

Evidently, if our arguments against information markets hold water, there is no, or at most highly fragmentary, supply of information in the strict sense. For one thing, information search should be substituted for price signals, which implies that quantities cannot be guaranteed; for another, the effectiveness of a firm's search is influenced by the quality of the information 'supply'. Another unusual characteristic of the supply of information is the suppliers' lack of economic motive: the 'sources' of information - individuals and organisations of the knowledge system - often feel no incentive to supply information to firms, a fact which is of course the counterpart of our assumption on an active search on the part of the firm. The conclusion could be that the supply of technological information is seldom 'custom-made' and, given the firm's information profile, often inaccessible.

As a first approximation, we define the quality of information supply as a combination of the scientific quality of the knowledge available in the knowledge system, its commercial applicability, and its accessibility in the sense of being explained in understandable language. From the information profile of a firm and the quality of the information supply, the innovativeness of a firm can be formalised as follows:

$$\underline{i}_j = a.\underline{ip}_j^{\alpha}.\underline{is}_j^{\beta} \qquad (1)$$

where:

$\underline{i}_j$ = innovativeness of firm j
$\underline{ip}_j$ = a vector of information-profile variables of j
$\underline{is}_j$ = a vector of information-supply variables of j

Some arguments could be given for the above specification; however, we are less concerned with the specification than with the empirical falsification of any innovation model. From the information paradox cited earlier, a firm does not know its information supply, which leaves us with a serious problem of empirical observation (unless a scientist does know the supply, which would make him a much wanted person among innovating firms ...). The analysis of a number of innovations aggravates the problem, because every innovation may have its own supply of information. Equation (1) tells us only how one innovation of firm j depends on the firm's profile and the quality of information supply for that particular innovation.

We have tried to 'solve' this problem by shifting our attention from knowledge as a supply of bits of information, each with a certain quality, to supply as the presence of organisations and individuals engaged in producing technological knowledge. The quality of supply is thus measured as the general (that is, not restricted to a specific piece of information) scientific standard, application orientation and business language of a specific organisation. In that way, the methodological problem of measuring the knowledge needed for each innovation, is avoided to some extent, because a 'supply organisation' can deliver information for certain kinds of information, for instance concerning a technological trajectory for one or a small number of industries; of course, even in that broader scope, the quality of a knowledge producer still has to be determined.

Equation (1) can be re-interpreted on those grounds:

$$\underline{i}_j = \dot{a}.\underline{ip}_j^{\dot{\alpha}}.\underline{is}_j^{\dot{\beta}} \qquad (1')$$

where:

$\underline{i}_j$ = innovativeness of firm j;
$\underline{ip}_j$ = vector of information-profile variables of j;
$\underline{is}_j$ = vector of qualities of technological information producers with regard to j.

Innovativeness could now be interpreted as the number of innovations realised by firm j (related, of course, to a time period, which has been left out for analytical simplicity), if those innovations are based on a knowledge input which can be traced back to the same set of knowledge producers (an identifiable subset of the knowledge system). That condition could be formalised as: $\exists(\underline{is}),\ j \in J \rightarrow \underline{is}_j = \underline{is},\ \forall\ j$, where J is the group of firms j which have a similar technological envrionment $\underline{is}$.

This general information behaviour of firm j enables us to extend the analysis to sectors.

FROM THE MICRO TO THE MESO LEVEL

We have already suggested that a knowledge-producing organisation could be relevant to a technically restricted set of innovations, for instance those created within a sector of activity that is homogeneous with respect to the technological knowledge required for its innovation processes. In practice, a sector is not always composed of firms producing the same type of product; the 'steel industry', for example, combines firms using similar process technologies; 'medical instruments' is a sector of highly heterogeneous establishments bound together to some extent by the 'destination' of their products, etc. For analytical clarity we therefore define a sector as: a group of firms with a homogeneous technology and/or a homogeneous product (or set of related products), depending for their technical innovations on the same set of knowledge producers. Obviously, this limits the analysis to the short run, because a firm cannot 'diversify' itself out of a sector of our description, diversification being constrained within the bounds of knowledge supply; to our mind, such a concept is comparable to that of a 'technological trajectory' or 'natural trajectory' (refs. 7-8), although our distinction between a firm's (exogenous) information supply and (endogenous) knowledge accumulation probably indicates a significant difference.

On our assumptions, the innovativeness of a sector can be formalised as:

$$\underline{i}_k = a^{*}.\underline{ip}_k^{\gamma}.\underline{is}_k^{\delta} \qquad (2)$$

where:

$\underline{i}_k$ = innovativeness of sector k;

$\underline{ip}_k$ = quality of the information profile of k (some average of firms j);

$\underline{is}_k$ = quality of knowledge producers related to k;

and:

$\underline{is}_j = \underline{is}_k, \; j \in k$

The innovativeness of sector k (the number of innovations per time period) could in principle be estimated from equation (2), which explains innovations as far as it depends on external information. Of course, innovation could be related to all kinds of other variable, such as availability of risk capital, the profit rate of sector k, the size distribution of firms j in k, etc. One aspect of information itself which has so far been excluded from or analytical treatment, is a medium between source and user. From theoretical as well as empirical research (refs. 9-10), it is becoming increasingly clear how much firms, especially smf, depend on intermediate organisations which select and transfer information form the scientific community and other 'potential' sources of technological knowledge to the production system. Translated in terms of our simple model, that phenomenon suggests that the quality of supply is too poor to ensure sufficient knowledge accumulation on the part of firms; in particular, the production of knowledge is inadequate in terms of applicability and the use of inaccessible theoretical language. We know moreover from the results of fragmentary empirical research that smf often lack a reasonable information profile ($\underline{ip}_j$).

Following the same line of thought as with respect to 'quality of supply', we define the quality of information transfer of sector k as the quality of organisations which - mostly on a non-commercial base - are occupied with transferring information from knowledge system to production system. Model (2) could be extended to

$$\underline{i}_k = a^* . \underline{ip}_k^{\gamma} . \underline{it}_k^{\delta} . \underline{is}_k^{\varepsilon} \qquad (3)$$

where:

$\underline{it}_k$ = quality of information-transfer organisation 'around' k.

($\underline{is}_k$, $\underline{it}_k$) could be called the meso-information structure of sector k. This meso-structure could be interpreted as a type of economic organisation somewhere between market and plan, which owes its development to a high degree of imperfectness of information markets with respect to technical innovations. We have found similar arguments for intermediate types of organisations between market and hierarchy - following Williamson's distinction - in Imai and Itami (ref. 11). To strengthen this intermediate structure could be part of a regional or national industrial policy. In Western Europe several elements of national innovation policies could be interpreted in that way, such as the Business Extension Scheme in the U.K. and the 'centres technologiques' in France (see ref. 1).

The fact that our analysis focuses on sectors rather than a group of complementary firms or even a group of related sectors (the 'filières' in France) might be an important weakness. The same question arises in the context of the estimation of K sectors; we will complete our analysis by some such estimation problems.

PROBLEMS OF EMPIRICAL OBSERVATION

One of the main obstacles to empirical research in innovation theory is the difficulty of measuring innovation itself; from the literature we are familiar with the choice of R&D expenditures (input) and patent registrations (output) as indicators of sectoral innovation. Both can be criticised on empirical and methodological grounds; moreover, data on an acceptable disaggregate level are often lacking (in the Dutch situation: the 4-digit level of the SBI). One possibility of improving measurement is to clarify the definition of innovation by introducing different types of technical innovation, thus making the number of achieved innovations measurable. On the other hand, such a distinction stresses even more the problem how to compare different innovations among firms or even within a firm; the determination of a sector's innovativeness by 'counting' the number of innovations could lead to the neglect of important sectoral differences.

With regard to our specific 'information' approach to sectoral innovation, serious problems have to be admitted with identifying and estimating 'information transfer'. Often neither the supply nor the transfer organisations are sector-specific, which means that it is not clear for which sectors a specific organisation might be relevant. This implies the problem of identifying a non-diagonal matrix with sectors on one side and knowledge-producing organisations (or transfer organisations) on the other. Such a matrix for information supply can be defined simply as: $M^s = \underline{s}.\underline{is}'$, where $\underline{s}$ is the vector of sectors k, and $\underline{is}$ is the vector of knowledge producers relevant to the production system (subsystem) $\underline{s}$. Similarly, $M^t = \underline{s}.\underline{it}'$ represents the matrix concerning transfer organisations $\underline{it}$.

The construction of such a matrix for a national economy (albeit restricted to the number and scale of organisations involved in the transfer of technical knowledge to the private sector) probably reveals sectoral connections which could give an insight into the limitations of a sectoral approach to innovation policy.

SUMMARY AND CONCLUSIONS

We have tried to explain the construction of a simple model of sectoral innovation which reduces the complex interrelated and dynamic processes of information accumulation to a static structure (information chain) containing the information supply, transfer capacity, and 'information-search-and-utilisation capacity' of a sector; different levels of sectoral innovation are thus related only to different information structures (different only in terms of values of the variables constituting the structure).

Such a model might be criticised on several grounds, some of which we will mention ourselves:

- the dynamic features of knowledge accumulation are far more important than the structure at a specific point in time;
- within a sector, firms may differ widely in terms of cost structures or strategic choices, which may influence the innovation process considerably, as far as susceptible to translation into an 'information profile', the variation of profile values within a sector is not taken into account;
- the measurement of sectoral innovativeness is subject to much uncertainty and to arbitrary choices which may not permit correlation with any 'explanatory' variable at all.

Nevertheless, if the importance of technical innovations in modern market economies is duly recognised, to widen the scope of empirical research is imperative. A static analysis of intersectoral differences has already given some insight into the possibilities and limitations of this kind of analysis. An important feature of our model is its link - simple as it is - to firm behaviour; extended in that direction, the model might be useful for political purposes, because it relates the improvement of the information infrastructure to the innovation performances of individual firms. Unfortunately, the present severely constrained data availability does not make such an extension very promising in terms of empirical falsification.

REFERENCES

1. P.R. Beye et al. (eds.), Towards a Competitive Future for Europe, Croom Helm, May, 1987.
2. U. Colombo, Innovation Strategies for Government & Industry, in: A. Gerstenfeldt (ed.), Technological Innovation: Government/Industry Cooperation, Wiley, New York, 1979.
3. M.V. Porat, The Information Economy, Stanford University Press, Stanford, 1976.
4. F. Machlup, Knowledge: its Creation, Distribution and Economic Significance, volume 3, Princeton University Press, New Jersey, 1984.
5. K.E. Boulding, Knowledge as a commodity, in: Beyond Economics, Essays on Society, Religion and Ethics, University of Michigan Press, Ann Arbor, 1968.
6. K.J. Arrow, The economics of information, Collected Papers of Kenneth J. Arrow, vol. 4, The Belknap Press of Harvard University Press, Cambridge (Mass.), 1984.
7. R.R. Nelson and S.G. Winter, An Evolutionary Theory of Economic Change, The

Belknap Press of Harvard University Press, Cambridge (Mass.), 1982.
8. G. Dosi, Technological paradigms and technological trajectories, Research Policy, vol. 11, no. 2, June 1982.
9. R.G. Havelock, Planning for Innovation through Dissemination and Utilization of Knowledge, Ann Arbor, Michigan, 1971.
10. NOBIN, Externe informatieverzorging van het kleine en middelgrote industriële bedrijf, Eindrapport van de Stuurgroep PBIB, The Hague, 1983 (External information provision of small and medium-sized industrial firms).

Innovation: Adaptation and Growth, edited by R. Rothwell and J. Bessant
Elsevier Science Publishers B.V., Amsterdam, 1987 — Printed in The Netherlands

HIGH-GROWTH TECHNOLOGY-BASED SMALL AND MEDIUM-SIZED FIRMS IN EUROPE

Mark Dodgson
Design and Innovation Management Group
Science Policy Research Unit
University of Sussex
United Kingdom

ABSTRACT

High-growth, technology-based small and medium-sized firms are becoming increasingly important means of technology generation and transfer and employment creation in Europe. While their creation and expansion have been encouraged by various recent structural and cultural changes, many serious constraints exist. By examining a number of highly successful firms, this paper examines how they overcame past problems and identifies future constraints facing them. This paper considers management, personnel and skills, finance, and technology in the firms, and emphasises the importance of the effective management of technology for their growth and success.

HIGH-TECH SMFS IN EUROPE

The proliferation of high-growth, technology-based small and medium-sized firms has been, until recently, primarily an US phenomenon. Increasingly, however, such firms are assuming a significance in European industry. A report by Arthur D. Little[1] in 1977 examining new technology-based firms found that there were only about 200 such firms in the UK, and probably even fewer in the Federal Republic of Germany. Ten years later, however, numbers have increased considerably. A study by Segal Quince and Wicksteed in the UK and the ISI in West Germany[2] estimated than in 1985 there were between 6000 and 7000 NTBFs in the UK, and around 3000 in the Federal Republic. In total employment terms, the 1977 study estimated there to be 27,000 employed in UK and FRG NTBFs; the 1986 study estimated this number to have increased to 240,000.

There are numerous reasons for this expansion in the number of new (and therefore almost entirely small and medium-sized) technology-based firms. Information technology, biotechnology and new materials technology are among a number of recent developments to have provided opportunities for the growth of less capital-intensive, more knowledge-intensive firms, with subsequent lower entry costs. New capital markets have been created, and there has been a considerable increase in the number of venture capital organisations. In Britain, for example, there are now 110 members of the British Venture Capital Association, and the European Venture Capital Association has over 160 members.

There has been a growing awareness on the part of national governments and of the EEC of the importance of high-tech SMFs, and consequently an increase in policy measures designed to encourage their formation and growth. The FRG, for example, under its Pilot Scheme for the Promotion of NTBFs, now offers grants as high as up to 90% of the costs of NTBF inception[3]. The EEC now has a Commissioner responsible for SMFs and a Task Force to examine the position and potential of SMFs, and to advise on how best to realise their potential.

These changes in policy have been matched by an important change in attitudes towards being more favourable to the small entrepreneur, and in the enhancing the image of small, high-tech companies. This is perhaps most marked within academia, where it is no longer considered unusual for a scientist or engineer to be entrepreneurial. The growth of science parks or innovation centres on campus is both a consequence of, and a contributor to, the change in attitudes in Universities. Whereas 15 years ago there were no science parks in Britain, there are now 24.

The above are amongst many factors to have contributed to the rapid growth of high-tech SMFs. Whether the present number of these firms is likely to expand at a similar rate in the future depends very much on the provision of suitable infrastructural and tangible assistance to such companies. This paper will, by examining a sample of highly successful, high-growth, technology intensive SMFs, attempt to identify some of the major problems faced by such firms, some of the reasons for their success, and what are seen as major constraints in the future.

THE RESEARCH

The Research on which this paper is based was conducted for the EEC's Industrial Research and Development Advisory Committee - Working Party III (concerned with SMFs). This paper will refer to the research conducted by the author as part of a larger study[4].

The Sample Firms

Twelve firms were visited during the course of this research; six British, three Danish, two Dutch and one Irish. The oldest firm was established in 1874, the youngest in 1981. Three firms were established before 1920, four between 1950 and 1970, and five after 1970. Between them the sample companies showed remarkable growth in their annual turnover between 1983 and 1985. The average (mean) rate of growth of each company during this period was 127%, and in two cases it was over 300%. Collectively, the 12 firms had annual sales of 212 million ECU in 1983. This increased to 261 million ECU in 1984 and 322 million ECU in 1985; a growth rate between 1983 and 1985 of 51.8%. Total employment grew by over 1000 between 1983 and 1986. It grew by 11.2% between 1983 and 1984; by 18.7% between 1983 and 1985; and by 30.7% between 1983 and 1986.

Data on profitability were incomplete. Two firms' accounting procedures did not reveal profit before tax, only after tax (because of complex tax procedures). Generally, profit levels were high. Excluding one company (which was only a few years old and had planned losses), average profit before tax as a percentage of turnover in 1985/86 was 13.3%. This figure does not include the two firms reporting average after tax profits of 8.7%. Three firms reported zero profit or losses in 1983: they were all recently established; two have now moved into profit, while the third is rapidly moving towards profitability.

The companies generally were highly export-oriented. Export ratios varied from 5% to 100% of sales, with an average export ratio of 68.7%. By far the most important overseas market was the USA. Seven companies exported to the USA where, in 1985, they exported goods and services worth 88.1 million ECU. The second most important markets for the firms were European.

TECHNOLOGY

The companies visited were all highly technologically advanced. Some were working at the 'leading edge' of their technology. All the companies were reliant largely on internally-generated knowledge, claiming that their in-house R&D activities are the major source of technical know-how and the main basis of their technological accumulation. Nevertheless they all enjoyed a variety of often very strong external technological links with Universities, and research institutes, and with other industrial companies.

Most of the sample firms initially gained technological expertise - and began their movement up the technological learning curve - on the basis of externally acquired know-how. External linkages were still very important: nine firms claimed important technological links with Universities; four claimed links with government research institutes; two used licenced-in technology, two have created joint ventures; and two made acquisitions in order to access external technology. One company belonged to an industrial club, while a second used the European Space Agency Database which it regarded as an important source of technical information. In Britain, four of the six companies were participants in the Alvey Programme of collaborative research into advanced information technology.

Three companies began their lives as University spin-offs, and while two no longer have strong technical links with their old University, they have forged strong links with research groups in other Universities. Two other companies began their lives dependent on technology from government institutes or departments, with which they retain very strong links. One company started as a joint venture initially obtained all its technological know-how from its US partner. Two firms gained their initial technical competence through licencing US technology.

The companies all devoted a considerable percentage of their resources, both manpower and financial, to their research and development activities. The actual R&D expenditure as a percentage of sales varied from 5% to 40%, with an average (mean) of 12.4%. (The data are for eleven firms; one firm declined to reveal its R&D expenditure.) These figures are for internally-generated R&D expenditure, and do not include the often large amounts of R&D undertaken for external concerns.

The high level of resources dedicated to R&D are matched by the high number of R&D personnel employed. Excluding one company which estimated that all its employees were involved in R&D, a total of 593 employees (or full-time equivalents) work on R&D in the firms; 17.2% of the total number employed. The average intensity of R&D employees/total employees within the eleven firms is 25.6%.

Although it is difficult to generalise with such a small sample, it is interesting that the companies with the highest proportional R&D expenditures and manpower intensities were the younger firms. The three firms which spend more than 15% of turnover on R&D, which were also the only firms with R&D manpower intensities of greater than 40%, were established after 1978. Equally of interest, and perhaps going some way towards explaining their continuing success, is the fact that all the firms created before 1920 dedicate at least 5% of turnover to R&D today, and have significant R&D intensities.

Among the most important continuing sources of technical information is feedback from the users and suppliers. All of the firms paid particular attention to feedback from customers, while eight firms specifically mentioned this source as being of key importance in the company's technological development. In a number of cases, user feedback was formalised; in one firm, for example, senior management paid annual visits to all main customers to discover how service could be improved. Links with suppliers were also extremely important, particularly in the electronics area, where firms made took great pains to identify new developments in microprocessors which they could incorporate in their products, and new advances in manufacturing methods which they could profitably adopt.

The general picture within the firms, therefore, was one of external technical expertise and internal knowledge generation being seen as necessary complements in the overall process of technological know-how accumulation.

MANAGEMENT

One of the major reasons for the success of the companies was their effective management. Often the most critical junctures in the companies' development (and a number had nearly gone bankrupt in the past) occurred through lack of managerial skill or through managerial complacency, and were overcome through changes in management. These critical junctures usually occured as the company was going through a transitional phase in its development; moving from its start-up phase, internationalising its markets, or seeking synergistic or broader diversification[5].

Many of the firms faced problems, for example, in making the transition from an entrepreneurial to the professionally-managed mode of operation. Founders of such companies, albeit imbued with the requisite technical and scientific skills combined with entrepreneurial ambition and flair, are not always best equipped to handle the more formal aspects of management, i.e. creating financial control and reporting systems, personnel management, creating effective organisational structures; nor are they much given to delegating responsibility. These are, however, all essential requisites for young companies wishing to grow rapidly.

Three companies were rescued from insolvency through changes in management; one company attributed much of its success to the early appointment of a talented manager; another company had to appoint a new manager to oversee dramatic changes brought about through rapid growth. A company formed as part of a public policy initiative attracted private sector funds mainly because of the exceptional quality of its founding management team; and one very small company gained considerably through their access to managerial and financial advice from its shareholding bank.

Almost all the chief executives interviewed had a good knowledge of the technology in which their firms operated, and eight possessed formal technical qualifications. They were all acutely aware of their companies' technical strengths, their competitors' technical strengths, and hence of their own comparative technical advantages. Most had formulated quite sophisticated stategic plans for technological development; they had identified future strategic market niches and were actively planning the accumulation of technological advantage needed to achieve a sufficient market share in them, either through internal R&D, licencing, joint venturing, by acquisition, and most commonly by employing a combination of these. Many perceived a future shortage of available entrepreneurial managerial talent; which reflects the concern that expansion would be limited by the lack of personnel capable of the dynamic leadership required to develop new products aimed at these new market niches.

PERSONNEL

The success of these companies, as explained by their CEOs, was based on the quality and commitment of their employees. The disadvantages the companies felt concerning their inability to match the salaries offered by large firms, was to some extent counterbalanced by the stimulating working environment of a small, high growth, high-tech company, and the opportunities for personal progression therein. It was recumbent upon the senior managers in these firms, therefore, to provide opportunities for personal and career development to satisfy individuals, while at the same time ensure that this benefitted and matched the needs of the firm. To attain this synthesis required considerable management skills.

Within the R&D function, the firms managed to attract a high number of graduate scientists and engineers. In ten of the firms, 270 out of the 489 employed in R&D were graduates or had postgraduate qualifications (55%). One of the remaining two firms, which claimed all its staff work on R&D, employed 200 graduates, while the other had 150 science graduates working in the firm "most" of whom work in R&D (it was impossible to obtain a precise figure for R&D employment within this firm).

The most widely articulated concern about future obstacles to growth related to shortages of key personnel. The most crucial personnel shortages mentioned as being a major constraint on future growth were: lack of entrepreneurial management (6 firms); shortage of electronic engineers (5 firms); and shortage of marketing managers (4 firms).

The shortages of electronic engineers generally arose through the firms' inability to offer salaries competitive with those offered by large companies. The shortage in marketing managers was most commonly mentioned by companies attempting to move into US markets. This was a particularly crucial constraint given the enormous US market potentials involved; in one case the Californian market alone was twice the size of the total European market.

FINANCE

While access to start-up finance was not a particular problem, funds to assist subsequent growth or to facilitate vertical integration were often extremely difficult to obtain, especially for the independent companies. Companies in particular experienced problems in obtaining finance for long-term R&D, to assist the move to manufacture, or to support major changes and expansion in manufacture. Again, the ability to raise finance for these technological developments required considerable managerial skill.

CONCLUSIONS

High growth, high-tech SMFs are an important component of advanced industrial structures. They can profitably generate and transfer technology in a number of key developing technologies, and can efficiently provide for the growing number of niche markets these technologies create. While not yet a significant job generator in total employment terms, they do create jobs and those jobs tend to be highly skilled. Nation state and EEC policies increasingly realise the importance of this type of firm. To understand some of the constraints facing the formation and growth of these firms a small sample of very successful high growth, high-tech SMFs, have been studied and certain lessons can be learnt from them:

-- The companies were very good at accessing and developing technology. They were generally founded on the basis of very strong external links, and their internal R&D commitment has been considerable. External technological links have been progressively broadened.

-- The companies that have experienced severe financial or market difficulties, did so largely because of poor or complacent management. The companies overcame these difficulties through the appointment of new and more skilled management.

-- The success of these companies is attributed by the CEOs to the skills and commitment of their employees. Attaining this commitment, and creating a stimulating working environment, requires a considerable effort on the part of senior managers.

-- As in the past, the future of these companies depends on the quality of management they can attract, and how effective they are in the management of technology.

Although the number of high growth, high-tech SMFs in Europe is increasing, their number are still relatively low compared to the USA. Numerous factors need to be taken into account when attempting to prescribe policies to increase the scale and scope of these firms, both infrastructurally and culturally.

Within the firms themselves, study of a small sample shows that one of the major reasons for their success is the effective management of technology. This requires accessing and developing technology by building extensive links with other research institutes, significant and appropriate R&D efforts, recruitment of skilled people and the creation of working environments where these skills are used and commitment and motivation engendered, identifying and satisfying suitable market niches, and the ability to raise finance for technological development often from sources unaware of technological matters and wary of funding young and small concerns.

One of the major distinguishing characteristics of the sample firms was the ability of senior managers to generate coherent strategies for technological development taking all these considerations into account. This research, therefore, points to the need to improve the ability of managers to think strategically about technology, and to access, develop, and market it effectively.

Whether indeed there are the mechanisms within Europe to train and retain managers in the management of technology is open to doubt. In Britain, at least, most managers lack formal management education and training[6], and in the case of high-tech SMFs 63% undertake no management training at all[7]. Given the potential of the high-tech sector in an Economic Community which has an unemployment level of over 12m, and imports 80% of its personal computers, this is a problem that requires immediate redress.

NOTES

The author would like to acknowledge his thanks to Professor Roy Rothwell for his contribution to and comments on this paper.

1 Arthur D. Little, New Technology-Based Firms in the United Kingdom and the Federal Republic of Germany, Anglo-German Foundation, London, 1976.
2 Segal, Quince and Wicksteed and ISI, New Technology Based Firms: A Review of Their Recent Performance and Significance, SQW, Cambridge, 1986.
3 M. Kulicke and H. Krupp, "The formation, relevance and public promotion of new technology-based firms", Technovation, 6, pp47-56, 1987.
4 IRDAC Working Party III, Study Report "SMEs and their role in the new technologies", EEC, 1987.
5 These 'phases of development' are discussed in the IRDAC report, op cit.
6 J. Constable and R McCormick, The Making of British Managers, British Institute of Management, London, 1987.
7 I. Mangham and M. Silver, Management Training - Context and Practice, Economic and Social Research Council, London, 1986.

Innovation: Adaptation and Growth, edited by R. Rothwell and J. Bessant
Elsevier Science Publishers B.V., Amsterdam, 1987 — Printed in The Netherlands

STRATEGIES FOR USER-PRODUCER INTERACTIONS

Dr. Brian Shaw

Polytechnic of Central London, Faculty of Management Studies, 35 Marylebone Road, London NW1 5LS (England)

SUMMARY

The transfer, exchange and pooling of existing and the pressure and threat of new technologies have resulted in an increased interdependence throughout the innovation, production and marketing cycles. These linkages reflect the development and creation of new relationships between producers, centres of excellence, intermediaries and users at the various steps in these cycles. This move to create more effective interaction between users and producers results from the need for a widened base of scientific and/or technological resources, increased flexibility, access to markets and risk sharing. In addition, in some industries access to 'state of the art' knowledge, which resides in the user, is crucial to the management of the innovation, production and marketing cycles. Also there is a need to establish a prior position in the complementary assets which facilitates the capture of innovation benefits for the innovator.

Strategies for user-producer interaction in the sample of thirty four medical equipment innovations studied were: (1) Communication Strategies; (2) User-Innovation Capture Strategies; (3) Window-on-Resources Strategies; (4) Complementary Assets Strategies; (5) Entrepreneurial User Strategy; and (6) Product Champion Strategies.

This research has added to the understanding and knowledge of possible innovation benefit capture and cost diffusion mechanisms and extended the bases for the locus of innovation benefit.

INTRODUCTION

The transfer, exchange and pooling of existing and the presence and threat of new technologies are impacting significantly at the strategic level. This has resulted in increased interdependence throughout the innovation, production and marketing cycles. These linkages reflect the development and creation of new relationships between producers, centres of excellence, intermediaries and users at the various stages in these cycles. This move to create more effective interaction between users and producers results from the need for a widened base of scientific and/or technological resources, increased flexibility, especially in the retention and management of multi-disciplined teams, access to markets and risk sharing. In addition, in some industries access to the 'state of the art' knowledge, which resides in the user, is crucial to the management of the innovation, production and marketing cycles. Teece (ref 1) also identified the need, in certain cases, for the innovating firm to establish a prior position in the complementary assets which facilitate the capture of innovation benefits for the innovator. For example, the commercialisation of a new drug such as gammanym, studied in Coleman et al (ref 2) requires the dissemination of information over a specialised information channel, i.e., doctors relied on the experience of peers, conveyed in interpersonal networks, for evaluative information about the innovation. Similarly, as Rogers and Shoemaker (ref 3) found, individuals are more receptive to innovations which are congruent with the value and interest they have acquired in the course of their professional socialisation.

INTERACTION STRATEGIES

Gardiner and Rothwell (ref 4) found that much of the impetus for further product development derived directly from farmers who demanded increasingly more powerful and versatile models. Howard Rotavators' strategy was to seek positively these design advances, indeed, at one stage, they operated a suggestion scheme for the farmers to elicit ideas for making improvements.

A strategy for forklift truck manufacturers, Senker (ref 5), was to engage with the user in the design of the user's materials handling system, the knowledge and co-operation thereby gained being an invaluable input into their own truck development programme.

As part of an interaction strategy, the timing of the linkage between the user and the producer is vital in effecting the successful development of the innovation. As Von Hippel (refs 6-8) found that the innovation process can be 'manufacturer-dominated' or 'user-dominated' and, especially with user process innovations, the user can control the timing of the transfer of the technology to the producer.

Von Hippel and Finkelstein (ref 9) found that two brands of automated analyser were more likely to be used in the development of new clinical test methods by users than others, i.e., Technicon and Abbott. The reason was that the design of the reagent system portion of the Du Pont auto analyser presented a major barrier to any non-Du Pont personnel, whereas no such barrier existed with Technicon and Abbott equipment. Thus, as Rothwell (ref 10) stated, producers should offer product designs that are user-flexible.

Studies of problem-solving behaviour by engineers (ref 11) show that problem solvers, when faced with a new problem tend to return to successful techniques that they have used previously. Von Hippel (ref 12) suggested, therefore, that suppliers of new technology need to acquire a 'window on need' by buying into an established company that specialises in an old technology of analagous functions.

Von Hippel (ref 13) considered that several interaction strategies could be used by product manufacturers in industries characterised by user-dominated innovation, for instance the setting up of: (1) User Groups; (2) Applications Laboratories - the manufacturer, through the application laboratory, provides free or low cost R & D help in working out the user application; (3) Custom Product Group - It is often advantageous for a first to the market manufacturer to offer good, fast, flexible help to that subset of users whose special needs have previously proved to foreshadow general demand.

Von Hippel (ref 14) found that computer companies recognised that customers develop some software packages that other users would find valuable. Some producers ignore these sources, others set up relatively passive and low-profile software exchange services for customers, whereas others, such as IBM, set up their Installed User Program Department (IUP), which co-ordinates the efforts of acquiring user-developed programmes.

Technicon, in search of commercially promising instruments that researchers at universities and other institutions might develop, offered the inducement of a research grant. In the scientific instrument industry, a scanning of scientific articles identified those who had reported their instrument innovations. Analysis of customer parts used also identified those who had lower use and, therefore, were potential user innovators who had developed something to reduce parts wastage costs.

A strategy analysed by Foxall and Tierney (ref 15-17) was that the user played an active entrepreneurial role in product innovation.

The LACE (Lab Awareness of Customer Environment) programme at Hewlett, described by Peters and Waterman (ref 18), had customers making presentations of their needs and reactions to Hewlett Packard products and services. In another programme, software engineers visited user sites to get direct customer advice.

The Czech Cotton Industry Research Institute (CIRI) made it a point of principle to be involved with a new technique from its inception to commercial launch. A number of the open-end development team were involved, together with the machine builders, in the market launch of the open end spinner, and in the dissemination of information and advice to users. (Rothwell ref 19). The mill, while being required to be a commercial operation, also served for experimentation and pilot plant trials in assisting the further development of the system. Further close co-operation between the Institute and a number of leading man-made fibre producers in Western Europe and North Africa had contributed to the search for and establishment of optimum conditions for open-end spinning of man-made fibres.

Townsend's research into the Anderton Shearer Loader (ref 20) illustrated close user-manufacturer interaction. The original design came from a small National Coal Board (NCB) engineering maintenance team. In the twenty five years following its introduction, the machine was continuously enhanced through a series of major and minor re-innovations totalling one thousand in all. The strategy of working from the strengths of both parties was illustrated where the NCB developed the control systems, for which it was equipped by virtue of its superior R & D facilities; the machinery manufacturers concentrated on developing the mechanics and power source of the machine, areas of their traditional experience. Additional strategies enhancing the close links developed between the NCB and the machine manufacturers were: frequent interchange of technical information and of personnel; an incentive award scheme to draw on the dispersed expertise in the NCB colliery workshops, which provided one third of all innovations in the ASL; and NCB's policy of licencing the design to overseas manufacturers, with reciprocal agreements on design advances.

From the outset of the project to develop the Fiberlase (Macdivitt ref 21), there was close collaboration between the technical design and development team at Pilkington, Barr and Stroud and surgeons at a major Glasgow teaching hospital. The interaction strategy developed here was the setting up of a joint working party composed of Barr and Stroud management and senior surgeons at the hospital.

Laage-Hellman (ref 22 and 23) described the strategy developed by ASEA in Sweden to get the market interested in the new ASP steel. The problem identified was that ASP's customers, i.e., the tool manufacturers, saw the longer tool life of the ASP tools forcing them to increase their market share to maintain the same capacity utilisation. This was in most cases a difficult task, since the tool makers in general had weak marketing organisations. ASP were, therefore forced to establish closer relationships with end-users. A three party co-operation, including end user, tool maker and ASP, developed and they agreed upon a test programme. The market introduction efforts were concentrated in selected priority application areas where the maximum results in terms of ASP tonnage and profit were.

Another successful joint research was the development of a method for powder injection into ladles. The corporate research was financed by MEFOS, made up of some thirty five steelworks, non-ferrous metal producers, mining companies and suppliers to these industries in Sweden, Norway and Finland. The initial four year development work was made at MEFOS, the pilot plant results from this moved to the steelworks. The strategy illustrated by this process innovation was the linking of ASEA's exceptional ability to design advanced electrical and mechanical equipment and the potential user's greater process and materials knowledge.

INTERACTION STRATEGIES IN THE UK MEDICAL EQUIPMENT INDUSTRY

The Sample

The sample of innovations studied in the research described here was from the medical equipment industry, there being thirty four innovations developed by eleven companies (ref 24). An example is the Exercise Test Monitor:

The Exercise Test Monitor

developed by P K Morgan Ltd, consists of high precision carbon dioxide and oxygen analysers. A null indicating XYT Recorder, a super-brain computer, a computer print-out, a ventilometer and a cardioscope. The software programming for the system represented approximately four man years of high quality work.

Functional Utility to the User

The monitor can measure any procedure involving ventilating and cardiac responses to exercise, examination and assessment of pulmonary and cardiac disease, exercise induced asthma and response to chemotherapy and the study of athletic training.

The prime characteristic identified from the study of the medical equipment innovation process is that there are multiple and continuous interactions between the user and the producer throughout the process. In the sample of thirty four innovations, twenty six (76 per cent) were developed through continuous user-producer interaction, resulting in twenty two (65 per cent) being successful, one too early to judge and three being failures. The failures were due to unsatisfactory technical performance. A successful innovation was one that achieved the technical and commercial objectives set for it. (ref 25)

Communication Strategies

Because the means of carrying out and the results of clinical trials are only immediately available to the consultant clinicians or physicians who do the trials, some means of dissemination of data to other potential users and producers must be effected. Therefore, strategies must be developed by the producer and the user to ensure effective dissemination and use of the data. The three principle means of communication were found to be: (1) personal contact through the worldwide networking system (41 per cent); (2) the presentation of papers at international conferences, professional meetings and industrial liaison groups (21 per cent); (3) publication of articles in professional journals (21 per cent). Because of the dominance of personal contact between scientists, engineers and clinicians within the very efficient worldwide informal and formal networking system, the nodes being the medical centres of excellence, it is essential that the producers develop strategies to lock into this network. International conferences are part of this networking system. The initial negotiation between companies and the presenters of papers regarding involvement in presenters' research and/or development is effected at these conferences. Similarly, the presenters of papers negotiate with potential industrial partners funding for research and development and possible joint development and commercialisation of user-innovations.

An example of this proactive strategy was that developed by Dr Paul Brankin of Oxford Medical Systems who visited users at least twice a week to discuss needs to source innovations and to monitor the performance of prototype and pre-production models of company equipment on test with the users. He considered it crucial that the flow of information regarding current and potential user needs was maintained, and that he created enough time to interpret the information being fed into the company. The company found that it was essential to have a dialogue with users on a continuous, long-term basis. This was best achieved via the use of hospital consultants. The company relied on close collaboration with the more junior consultants, and constantly sought out the up and coming doctors. The monitoring of potential was effected through continuous scanning of the literature, presence at all conferences on subjects of interest to the company, and word of mouth through the company's established grapevine of contacts in hospitals. When contemplating developing a new piece of equipment, the company was concerned that the problem that the equipment was being built to solve was of such a generalised nature as to create a viable market. Oxford's cadre of consultants, normally six, tended to feed to the company information on clinical problems that had general applications to a wider group of patients. This close continuous relationship was necessary because Oxford could not identify clinical problems, the 'state of the art' knowledge being with the consultants. Further, the company could not sell a piece of equipment without a favourable response to the clinical trials by the consultant.

The communication of the Ambulatory EEG Monitor illustrates the effectiveness of this strategy, in that four hundred delegates at the meeting of epilepsy and neurology specialists in Amsterdam in 1977 showed great interest in the results of the clinical trials of the Monitor. The company also heard of other work being done at the National Hospital for Nervous Diseases on integrated circuit amplifiers which could be attached to the head. This design was included in the Monitor, tests carried out in Oxford and the National Hospital, and papers presented on the tests in Vancouver in 1978. The favourable and rapid response to the paper resulted in the company manufacturing twenty machines in September 1978, which were sold before manufacture. Sales revenue rose to £30,000 in 1978-79, £446,000 in 1979-80 and £1m. in 1980-81. The clinician from the hospital in Oxford and the engineer from the National Hospital gave many papers on their work and these resulted in at least half the sales in 1978-79. In addition, the company persuaded its users to present papers to the two yearly seminar on ambulatory monitoring. The response was such that a whole day was given over to the presentation, as against the normal half day per subject. The further response, when the papers were published was very important for the company, and copies of all the papers were made available to potential users as part of the sales literature.

Further illustration of a proactive communication strategy was the seeking out of Dr Simon Godfrey, Professor of Paediatrics at Haddass University, Israel by Philip Morgan, the managing director of P K Morgan Ltd. There were two reasons for this: the first was that Philip Morgan saw a major growth area in equipment for paediatrics; the second was that Dr Godfrey was widely considered a brilliant consultant and was becoming a world leader in his field. Most of the new diagnostic areas identified by Dr Godfrey, Morgan saw would only be feasible with increases in the present level of knowledge and technology. To ensure that Dr Godfrey favoured the company and that the company became known as being interested in developing medical equipment for paediatrics, the company funded an international conference for about forty to fifty paediatric pulmonary physiologists from all over the world at Haddass University. The purpose of the conference was for these physiologists to hear of the research being done over the previous three to four years by Dr Godfrey and his colleagues, thus consolidating the network and creating a relationship between the company and the delegates. He also showed off the Morgan Exercise Test Monitor, which he had previously sold to Dr Godfrey.

User-Innovation Capture Strategies

As stated, the 'state of the art' clinical and diagnostic knowledge resides in the user (the hospital/medical school consultant) and the user has a high probability of deriving 'output-embodied' benefit (ref 26) from the innovation. The output-embodied benefits are, for example, the medical equipment developed is capable of achieving a particular clinical objective as, for example that identified for the Exercise Test Monitor in "The Sample" above. The achievements of the clinical objective gives the clinician a greater understanding of bodily functions, greater diagnostic ability and a basis for further research. The experience of testing and evaluating the equipment before it becomes commercially available gives the clinician a significant advantage both in terms of research expertise, and practice in the implementation of the therapy or diagnosis developed. In addition, the world-wide diffusion of knowledge, best practice and social benefits through the 'Invisible College' (ref 27) increases the status of the consultant in this college. The closer these benefits are to 'state of the art' advances, the more involved the consultant will want to be. It is, therefore, essential that the producer searches proactively for user innovations and develops jointly with the user these and other innovations so as to ensure for the user these output-embodied benefits. Illustrations of proactive strategies are:

Pickering continuously monitors papers presented at international conferences and professional institutes, e.g., Institute of Hospital Physicists, Intensive Care Society, British Institute of Radiology, Physicists and Radiologists, and the journals, e.g., Radiography Journal. An article by J White of Kings Lynn Hospital on miniaturisation of radiography resulted in a major joint research programme.

Johnson and Johnson commissioned one of their senior personnel for one year to analyse and evaluate the content of all the scientific and medical papers that could be of concern to the company's agreed areas of interest. In addition, he monitored the predictions of the results arising from the papers of each author whom he considered to be worth following. On this basis, the company developed a Critical Care Monitoring Division by buying up companies that had previously been set up to manufacture products arising from the research expressed in these papers. The director also approached all authors to elicit their response to joining the R & D team of the company, or linking into the company's research programme in some way.

Philip Morgan of P K Morgan Ltd., when he introduced diffusion measurement and pulmonary function testing as major diagnostic techniques through developing and marketing the Transfer Test Apparatus, acted as an 'honest broker' and 'reference point' for pulmonary function testing. The chest physicians who were not familiar with these new techniques were loathe to ask colleagues about the results of their tests, which they did not fully understand, since they would consider this a loss of face. The result of this was that they were encouraged to contact somebody who could give them the answers, but was also not a threat, Philip Morgan. They forwarded their test documentation to him and he provided interpretation by return. If he was not absolutely sure, he would approach one of his chest consultant contacts, who was using the company's equipment, and ask them to interpret the results. This they were only too willing to do as it gave them added data for their own research work. This impartiality and expertise and close and continuous interaction with users resulted in sales of some 2,000 units at approximately £3,000 each up to 1981.

These strategies result in the fullest co-operation from the user, creating good internal and external communications and a full understanding of user need, the two factors which research (ref 28) has found to be significant determinants of innovation success. In this sample of thirty four innovations, eighteen were 'user-dominated', where user prototypes were transferred to the producer, and in another eleven, the user transferred the idea. The effectiveness of these strategies is illustrated in the success rate detailed "The Sample" above.

'Window on Resources' Strategies

The existence of the intermediaries, i.e., the Medical Research Council (MRC), Department of Health and Social Security (DHSS), Department of Industry (DoI) and the British Technology Group (BTG) in the networking system (ref 29) created a window on resources for both the producer and the user. The user and producers developed strategies to: (1) capture resources in terms of MRC engineering and scientific personnel being seconded to units to help develop user innovations; (2) use the expertise of the DHSS consultants to place prototypes in hospitals and medical schools for clinical trials, to evaluate finished products and purchase prototypes and final products; (3) to obtain BTG patenting - eleven of the pieces of equipment were patented, of which ten were successful; (4) to gain bridging finance from the DoI throughthe purchase of pre-production models by the DoI. The success of these strategies was such that the producers' R & D was supplemented by that of thirteen major research and professional institutes.

Complementary Assets Strategy

The very effective networking system described above, which creates strong linkages to sources of technology and information external to the centres of excellence, is a major source of consultants' power in these centres. The dominance of the consultant in the process of effective use and purchase of medical equipment results from their perception of the effectiveness of the equipment in improving diagnostic skills or enhancing therapeutic ability. The prime basis for these judgements is the clinical results achieved by using the equipment. The acceptance of these results is a function of who carried out the clinical trials and their degree of personal contact in the network. This 'peer group' buyer behaviour relayed very specific signals to the medical technology decision makers, the Area Health Authorities, and were fully accepted. Evidence was present in twenty seven of the innovations of peer group buyer behaviour. The strategies to lock into this complementary asset were for instance, to fund university chairs, funding of consultants' research, the loaning of equipment, providing free of charge or at cost consummables, funding consultants' visits to researchers in other parts of the world and to international conferences and the payment of retainers to them. (Eight of the companies studied paid these retainers.)

Entrepreneurial User Strategy

Three of the companies studied were set up by entrepreneur users to develop further and to market products which they considered warranted the risk of starting their own business, i.e., Innotron Ltd., P K Morgan Ltd., and Oxford Instruments. Two other companies studied, Rigel Research Ltd and Surgical Equipment Supplies Ltd were also set up by entrepreneurs who worked in the Medical Equipment Industry, Rigel being set up from scratch and Surgical being a management buyout by the managing director. They identified gaps in the marketplace and developed and marketed the innovations described in this study. Innotron was started by Professor Chard of St Bartholomew's Hospital to develop a Multiple Detector Head Gamma-counter for Radioimmunoassay. Professor Chard was an expert in and a founder member of the study of radioimmunoassay. The first multi-detector system was developed by Searle Analytic, but was superceded by the sixteen head detector manual counter, the NE1600, developed by Nuclear Enterprises Ltd. This counter reduced the total counting time from 85.3 minutes to 3.3 minutes (ref 30). Professor Chard was Nuclear's consultant on radioimmunoassay and it was as a result of a brainstorming session which he led that the concept of the NE1600 was born. Because of its quite dramatic impact on the productivity of the routine RIA laboratory, especially his own at Barts, Professor Chard decided to set up his own company to develop a solid state gamma counter using sophisticated software so as to engineer around Nuclear Enterprises' patents. He was successful after two and a half years' development, extensive testing and evaluation and, by August 1981, fifteen months from launch, had sold 160 units of the Hydragamme 16, at a price of approximately £9,600 each.

Product Champions Strategies

The research concerning the role of the product champion in the innovation process has been principally concerned with the role of the champion *within* the innovating company. The evidence from this sample suggests the need to identify, in addition, the product champions *outside* the company or innovating organisation. These 'outside' champions, in co-operation with the 'inside product champions' determined the role of and the nature of the product development and the final form of the innovation. These 'outside' product champions, both users and intermediaries, were coupled with the inside product champions in all but three of the innovations studied. The strategy which seemed to be appropriate to increase the effectiveness of the innovation process was the matching of the status of the inside and outside product champions (ref 24).

DISCUSSION

These user-producer interaction strategies studied here have added to the understanding and knowledge of possible innovation benefit capture and cost diffusion mechanisms. More research should be carried out to determine the nature of these mechanisms which are so effective in increasing the probability of innovation success. Also knowledge of user-producer interactions could be developed further.

This study also identifies that the locus of innovation benefit is determined also by the nature and levels of co-operation existing among the parties to the innovation process, the relative closeness of user requirements to the 'state of the art' and the perceived non-embodied and output-embodied benefits arising from this co-operation. Where the co-operation is close, intense, multi-purpose and continuous, the locus is determined by the joint and several decisions to innovate which arise from strategies generated by all parties involved. Further studies need to be carried out, in other industries, to determine the generality of this finding.

REFERENCES

1 D J Teece, Profiting from technological innovation: Implications for integration, collaboration, licensing and public policy, Research Policy 5, 1986.
2 J S Coleman, E Katz and H Menzel, Medical Innovation: A Diffusion Study, Bobbs-Merrill, Indianapolis, 1966.
3 E M Rogers and F Shoemaker, Communications of Innovations, Free Press, New York, 1971.
4 P Gardiner and R Rothwell, Tough Customers: Good Designs, Design Studies, Vol No 1, 1985.
5 P Senker, Forklift Trucks, in: K Pavitt (ed) Technological Innovation and British Economic Performance, Macmillan, London 1980.
6 E Von Hippel, The Dominant Role of Users in the Scientific Instrument Innovation Process, Research Policy Vol 5, No 3 1976.
7 E Von Hippel, Transferring Process Equipment Innovations from User-Innovators to Equipment Manufacturing Firms, R & D Management, 8, 1, 1977.
8 E Von Hippel, Users as Innovators, Technology Review, Vol 80, No 3, Jan 1978.
9 E Von Hippel and S N Finkelstein, Analysis of Innovation in Automated Clinical Chemistry Analysers, Science and Public Policy, February 1979.
10 R Rothwell, Innovation and Re-innovation: A Role of the User, Journal of Marketing Management 2, No 2 1986.
11 T J Allen and D G Marquis, Positive and Negative Biasing Sets: The Effects of Prior Experience on Research Performance, IEEE Transactions in Engineering Management, December 1964.
12 E Von Hippel, Increasing Innovators' Returns from Innovation: Working Paper Sloan School of Management, MIT Cambridge MA, 1981.
13 E Von Hippel, Successful Industrial Products from Customer Ideas, Journal of Marketing, American Marketing Association 1978 (b).
14 E Von Hippel, Get New Products from Customers, Harvard Business Review, March-April 1982.
15 G Foxall and J D Tierney, From CAP 1 to CAP 2: User-Initiated Innovation from the User's Point of View, Management Decision, Vol 22, 5, 1984.
16 G Foxall, F S Murphy and J D Tierney, Market Development in Practice - A Case Study of User-Initiated Product Innovation, Journal of Marketing Management Vol 1 1985.
17 G Foxall, A Conceptual Extension of the Customer-Active Paradigm, Techno-vation, 4, 1986.
18 T Peters and R Waterman, In Search of Excellence, Harper and Row, New York, 1983.
19 R Rothwell, Innovation in Textile Machinery: The Czechoslovak Experience, Textile Institute and Industry, December 1977 (a).
20 N E D O, Innovation in the UK, N E D O, 1982.
21 H Macdivitt, The Light of Recent Development, The Business Graduate, Spring 1982.
22 J Laage-Hellman, Technical Development in Industrial Markets, Workshop on Innovation Strategy in Established Firms, Institute for Advanced Studies in Management, Brussels, 1984 (a).
23 J Laage-Hellman, The Role of External Technical Exchange in R & D. An Empirical Study of the Swedish Special Steel Industry, Research Report No 18 Marketing Technology Centre, Stockholm 1984 (b).
24 B Shaw, The Role of the Interaction Between the Manufacturer and the User in the Technological Innovation Process, D Phil Thesis, Science Policy Research Unit, University of Sussex, April 1986.
25 B Shaw, The Role of the Interaction between the User and the Manufacturer in Medical Equipment Innovation, R & D Management, Vol 15, No 4, October 1985.
26 E Von Hippel, The Dominant Role of the User in Semiconductor and Electronic Subassembly Process Innovation, IEEE Transactions on Engineering Management, Vol EM 24, No. 2, May.

27 D Crane, Invisible Colleges: Diffusion of Knowledge in Scientific Communities The University of Chicago Press 1972.
28 R Rothwell, The Characteristics of Successful Innovation and Technically Progressive Firms (with some comment on Innovation Research) R & D Management 7, 3, 1977 (b).
29 B Shaw, Appropriation and Transfer of Innovation Benefit in the UK Medical Equipment Industry, Technovation 4, 1986.
30 T Chard, New Developments in the Field of Gamma-Counting, Antibiotics Chemother, Vol 26, Karger, Basel, 1979

Innovation: Adaptation and Growth, edited by R. Rothwell and J. Bessant
Elsevier Science Publishers B.V., Amsterdam, 1987 — Printed in The Netherlands

CONDITIONS FOR THE TECHNOLOGICAL CATCHING UP

Matjaz MULEJ, School of Business Economics, University of Maribor, Yugoslavia (P.O.Box 180, 62000 Maribor, Yugoslavia)

SUMMARY

The countries, which are latecomers to the industrialisation process (like my Yugoslavia), need the technological catching up, or their (our) path leads us to the African Sahara standard of living. Not only investment, but people's knowledge and motivation can help a lot. For many coworkers to be activated, innovation of the managerial style is a precondition, going toward WE ALL THINK, WE ALL WORK. Not only the Suggestion System, but methods of Creative Cooperation are necessary, and several of them have been produced and introduced. We could introduce them only in the companies with an innovative culture and policy. So, ten preconditions for the entire business to become innovative, have been created. They are briefed here.

0. PROBLEM AND VIEWPOINT STATEMENT

In the complex international economic situation of today, only those countries, companies and persons can have a relatively good standard of living, which are able to compete successfully. The sources of such a capability are no longer the ownership of land, or of raw materials, or of capital, but more and more the human rresources with creativity (as property and activity), which leads toward good productivity, reasonable cost and good quality. Who is not able to do so, becomes/remains a donnator of presents to the more competitive, i.e. the more innovative ones, and his own path takes him more and more into the African Sahara type standard of living.

So, such a country/company/person/people has an urgent need for the technological catching up. Yugoslavia is such a case, too, and we tried to produce a model for it. Its general part will be presented here.

1. DEVELOPMENT PHASES DURING THE 20TH CENTURY

The differences started emerging and growing far earlier, since the West European industrialisation process had started, but it is the20th century, which made them so essential for the entire world. It is so not only due to its culmination of the problems in two World Wars and two World Economic Cryses, but also due to its creation of two very different vitious circles. This development can be sketched as follows (1):

Development phases of prevailing technology	Labour intensive	Capital intensive	Science intensive
Education (% of univer sity degrees in labour)	- 2%	2-4%	7-15%
Pyramide of knowledge			
Price of job in $ 1.000	2	4	6
R&D funds in % of GNP	- 2%	4-6%	10-12%
NYpc in $ 1.000 up to	2-3	5-6	12-15
Typical source of knowledge	Experience		Research
Managerial style invent ed in the same period	Traditional (Taylor's Sci entific mgt)	Human Relations	Human Resources
Innovators	Professional elite only	Professional elite and some workers (by Suggestion S.)	Professio. elite in more and more of creative cooperati on with all others

We have all been educated in a paradigm that the Tayloristic Scientific Management is the best possible. Though, in a lot of modern literature it is called "traditional", i.e. obsolete (2). Namelly, as we found, it causes the following vitious circle for the latecomers, who did not abandone it:

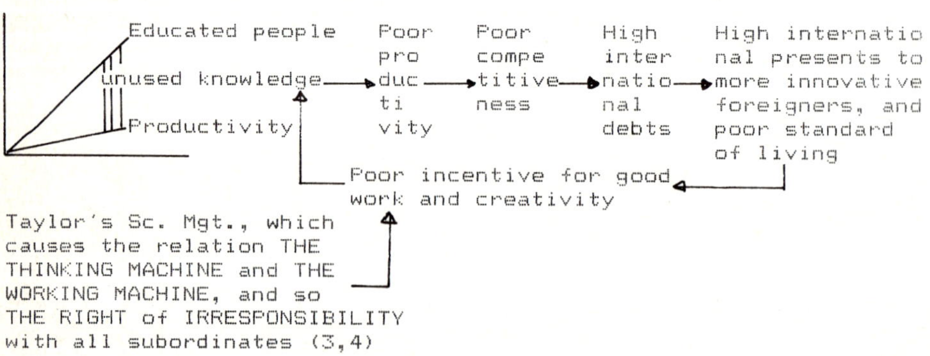

So, the conclusion is: only those companies and nations who have been able to move toward the modern relations WE ALL THINK AND WE ALL WORK, are able to be competitive, since they have the least of the unused knowledge and motivation of people, who become their associates, coworkers instead of being only workers (4). The ones who keep the old way, become more and more donnators of presents to the more innovative ones. So, they run more and more into the African Sahara type of standard of living, and they cannot afford education, R&D, investment or standard of living comparable with the ones at the top of innovativeness.

2. INNOVATION OF THE MANAGERIAL STYLE - THE FIRST PRECONDITION FOR TECHNOLOGICAL AND ORGANISATIONAL INNOVATION, AS A CONDITION FOR THE TECHNOLOGICAL CATCHING UP

In the course of the cited development, the number of educated persons has grown tremendously: in Taylor´s times there were so many teachers in the primary education around the world, as now are professors at the universities (5). If some managers, due to their education, still think that it makes no sense to ask subordinates any question, "because they cannot think", I ask them: "HOW COME THEY ARE ABLE TO SURVIVE WITH THE POOR WAGES THEY ARE PAID, IF THEY CANNOT THINK?"

They, obviously, can think. Their capabilities can be mobilised, if the managerial style is changed from the commanding one to an cooperative one (not only in Japan and USA, but also in a latecomer country, our experience has proven). What can the many subordinates be asked about, so to change from workers to coworkers? This is clarified with the finding that more than two thirds of innovation are small improvements about the own daily work procedure of the doers. Financially, they are less important, but they lead to the following chain:

COOPERATIVE MANAGEMENT
(WE ALL THINK, WE ALL WORK)

↓

(important innovation) and small improvement, contributed by many doers

↓

innovation friendly social climate in company/society

↓

understanding of and support to professional R&D work, which is the most essential source of innovation

↓

improved productivity, lower cost, better quality

↓

more competitiveness, and more presents received than given in the market

↓

better life and better motivation for good and creative work of many

↓

For such an innovation of the managerial style, a modernisation of mentality of the managers and of subordinates is an urgent necessity. In the most industrialised countries this process has had one or two centuries time to evolve. If a latecomer, like Yugoslavia (and som 140 more countries of 165 existing in the world), decides to wait for the natural evolution, the African Sahara standard of living will come too soon.

Something must be done earlier. After 1980, we started studying and developing methodologies of creative cooperation, based on the previously done theoretical work, for innovation to be produced by many coworkers, not only the professional elite (which, of course, remains the most important all the time, but changes from a subsystem into a partial system role). In Slovenia, the most western part of Yugoslavia (where I live and work, too), at least the following methods have been introduced (after 198o):
- Quality Control Circles, used for improvement to be produced by coworkers in the same department, be it on shop floor or elsewhere
- (my) USOMID. used for improvement to be produced on the basis of studying the entire process and its organisation, so to see interdependencies beyond single departments
- Value Analysis, used for improvement of products, mostly on shop floor
- Systems Engineering, used for solving more demanding problems by creative cooperation of higherly educated coworkers
- Business Improvement Projects, used for studying the general business problems of a project type
- Circles of Probationers, for the young newcomers' brains not to go routinistic and roasted, but creative
- CUM, used for many coworkers to engage themselves in search for new ideas to become the basis for new objectives of the company on its products and services program.

Soon, we all found, that the methods are accepted only by those companies, who already have had some innovative rather than routinistic orientation in their business and development policy and practice. We also found, that the market selection forces have not been working strongly enough in Yugoslavia, to produce an essential interest in innovation by themselves (6). So, the further research produced a book, written by me and seventeen coauthors, which is about to be published under the title "INNOVATIVE BUSINESS" (in Slovene). In autumn 1986, we could specify ten conditions to be fulfilled by a company, which finds it necessary to activate coworkers for the said chain, and so to break the vitious circle of the path toward the African Sahara standard of living.

By (uncomplete, but acceptable) data, namelly, the value added by innovation in Slovenia, and even more so in other parts of Yugoslavia, is very poor. But in companies, which follow an "innovative business" style of management, it has reached ten and more percent of total sales in 1985 in 1986 (7). With such companies, data include the R&D results and the amateur innovators results, as two partial systems of the same system, which support each other. Such figures, of course, are of big business importance, and they are worth the managerial effort, although the data, which include only the results of Suggestion System - are not worth it.

So, the old method of activating workers, limited to the Suggestion System (with a rather passive managerial style about innovation) as a system of random individual suggestions, needs a completion with methods of creative cooperation. And they need support with the following ten conditions for innovation (as activity, not only property), which involves so many coworkers so permanently, that one can stop speaking of the "invention and innovation activity" and can start better speaking of an INNOVATIVE BUSINESS as a wholistic concept of innovation and modern business.

3. TEN CONDITIONS OF INNOVATIVE BUSINESS AS A WHOLISTIC CONCEPT OF INNOVATION AND MODERN BUSINESS

The Suggestion System is, according to our experience, able to activate only fanatics of innovation. In a country, which is an industrial latecomer and so too full of routinistic traditional concepts, innovators are rare birds (white crows). So, they are in trouble and mostly give up. That is why the cooperative methods are necessary, among other reasons. With them, the results have been quite encouraging, but only if:
(1) the managers were very active themselves,
(2) they had a good Innovation Office, made of clever, ambitious and hard working coworkers.
But, in general, we must admit, that so far the theoretical knowledge has progressed much more than the practical results, although the applying companies have had good results, with virtually no exception.

So we defined the following ten conditions for business to become more wholistically innovative:
(1) innovation friendly culture and policy (of the government and companies management) for innovation to be supported and have an orientation,
(2) management by innovation objectives, for the top and middle managers to be motivated for innovation,
(3) education and training for innovation (as property and activity) and for creative cooperation of many coworkers (as a condition for many to be able to be involved in creation of innovation),
(4) organisation and running of invention-innovation processes - with a specialised office, Innovators Association in the company etc.,
(5) asuring of access to equipment, so the invention can be tested and developed into innovation, as well as access to professional and financial support and other condition for practice of innovation,
(6) assuring of information resources in form of literature, conferences, seminars, courses, professional fairs etc. (including the obligation of their participants to transfer the learned knowledge to their coworkers soon after return), for foreign knowledge to be applied, too,
(7) introduction of innovation into the regular work process so to avoid routinism of the old habits to prevail too much,
(8) marketing of innovation, and therefore a close cooperation of R&D and Marketing people, which should start early enough and goes on permanently (8),
(9) evaluation and rewarding of invention and innovation (with bylaws about it), in our practice with its basis in Suggestion System only until 1981 and with its prevailance even today, this one condition used to be regarded as the only one important, which it is only for fanatics, but they are not enough to solve the problem of technological cathing up of the country or a company,

(10) the legal and organisational set-up, which preferes innovation to blind (!!!) subordination, in order to create circumstances which are innovation friendly and so can activate the experience and thinking of (co)workers.

4. CONCLUSIONS

The technological catching up is the precondition for survival to all countries which are latecomers to the industrialisation process of the world. Without this catching up, they/we will remain (neo)colonial sources of free presents to the more innovative parts of the world, in any normal market business. Closing the borders (in autarchy) does not help, but causes an additionsal difference in creativity and innovation, because it cuts the sources of knowledge and the incentives, provided for by the market competition pressure.

The technological catching up cannot be achieved by investment only. If it takes place with a lack of own knowledge, the gap of productivity, cost and quality cannot be bridged over. The human resources must be activated. For this end, the managerial style needs innovation, first of all, so the traditional paradigm of the thinking machine and the working machine be abandoned, and will make room to the modern one WE ALL THINK, WE ALL WORK.

For this end, the traditional Suggestion System must no longer be the only or prevailing methodological basis to activate and motivate the brains of the Many, but the creative cooperation methods. They are able to activate many more coworkers, to change them from workers to coworkers and associates in an UGLY DUCK PROCESS, in which the innovators change from hated rare birds to beloved normal ones. This is a precondition for the R&D processes, the most essential partial system of the innovation process, to be well understood and supported by other members of the company's body.

For the innovation process, be it in the professional or amateur partial system, to be able to help us with the necessary technological and existential (!) cathing up, the transition from the concept, that the innovation is just one of many activities in the company, to the concept, that the entire business is an innovative one, is necessary. The alternative to this demanding transition is even more demanding: life in an African Sahara standard of living. I see no third way.

For this transition to be possible, ten conditions must be fulfilled, which are briefly described in this paper and elaborated more in our recent book and proceedings, partly they are still under an ongoing research (9).

REFERENCES:

1. This was found partly in the research project "Slovenia 2000", published in 1985 in Slovene, partly by me.

2. J. Zupanov, in J. Jerovsek et al., Crysis, Blockades and Perspertives, Globus, Zagreb, 1986 (in Serbocroate)

3. J. Matejko, the author of "In Search of New Organizational Paradigm", Praeger, 1986, said in a talk, given in steelworks Zelezarna Ravne in Slovenia, in March 1987, that every hired worker signs two contracts: an economic one and a psychological one, an nobody offers more of his/her capabilities than what he/she feels that his/her boss explicitly expects in reality

4. Let us just recall the classics, e.g. Peters and Waterman Jr., In Search of Excellence, R. Pascale and A. Athos, The Art of Japanese Management, J. Merkle, Management and Ideology, The Legacy of International Scienific Management Movement, R. Cantor, Change Masters, several cases in Inc. Magazine and other media, the Study of the New York Stock Exchange Office of Economic Research, People and Productivity, A Challange to Corporate America, W. Goldsmith and D. Clutterbuck, The Winning Streak etc.

5. M. Pecujlic, University of the Future, Nin, Beograd, 1980 (in Serbocroate)

6. All atempts of Yugoslavia to introduce fully working market laws have failed, partly at least, in the battle of the interests for the state redistribution and the interests for the market redistribution, which has lasted all four decades after WW II witout interruption. The new government, elected in 1986, speaks of the "real economy" again. This is important, but already outside the topic of this conference.

7. Data of Social Accountancy Service of Slovenia, and of enterprises TAM Maribor, Krka Novo mesto, Mariborska livarna Maribor and some others.

8. The ISPIM Conference in Innsbruck 1985 was my first good experience of R&D&M. By official data, published in Feb. 1987 by the Federal Council of Trade Unions of Yugoslavia, less than five percent of firms in Yugoslavia practice R&D regularly. The same is the percentage of the firms who practice the M = marketing approach to business, according to research, done by the Marketing Dept. of my school a few years ago.

9. M. Mulej and coauthors. Innovative Business, Gospodarski vestnik, Ljubljana, 1987 (in Slovene), M. Mulej (ed.), Proceedings of the conferences on the occasion of the Innovators Day in Maribor of 1981, 1982, 1983, 1984, 1985, 1986, published always in the review Nase gospodarstvo (with English abstracts) Maribor, always Nr. 2 or 3 or 2-3; there are manuals to nearly all of the cited methods of creative cooperation, and several other proceedings of conferences and books, all in Slovene or Serbocroate languages: the titles can be obtained by request from me.

Innovation: Adaptation and Growth, edited by R. Rothwell and J. Bessant
Elsevier Science Publishers B.V., Amsterdam, 1987 — Printed in The Netherlands

PLADOYER FOR THE USE OF QUALITATIVE METHODS FOR EVALUATING INNOVATION POLICY

S. MICHALLIK

Institut für Unternehmungsführung (Institute for Business Management), Freie Universität Berlin, WE 6, FB10, Garystrasse 21, 1000 Berlin 33, West Germany

SUMMARY

As in other european countries, the importance of innovation policy has increased steadily in West Germany over the past ten years. Parallel to expenditure on innovation measures, the need has also grown to evaluate the policy behind them. Studies were commissioned among a very varied range of scientific disciplines which pointed to the key role of attitudes inside the promoted firms. The outlook of management and other personnel on innovation and innovation policy is of vital importance for its success or failure, while being extremely difficult to pin-point. We plead in the defence of a more systematic use of qualitative methods, and have begun a project to demonstrate means of conducting practicable surveys on attitudes towards innovation and innovation promotion.

INTRODUCTION

From the mid-seventies onwards, the West German government started implementing innovation promotion measures to increase the international competitiveness of german industry. A whole package of measures was designed to turn new technology into marketable products in the shortest possible time.

As innovation policy is such a new science, little tested knowledge is as yet available about the workings of the various - often expensive - measures. Expenditure on promoting research and development (R&D) and innovation measures in small and medium sized enterprises (SMEs) rose from 131 million to 824 million DM between 1975 and 1982 (ref. 1). The range of incentives is wide (ref. 2):

- Indirect and indirect specific measures to promote R&D and innovation in firms (e.g. incentives to hire new R&D personnel, promotion of R&D investment or special technology i.e. micro electronics, biotechnology, CAD/CAM)
- Technology orientated project promotion (e.g. ocean technology, space research, transport & communications systems, ecology research, etc.)
- Business foundation grants, capital and credit loans, venture capital
- Technology transfer from Research institutions
- Technology and innovation counselling
- Organization of databases with scientific and other information

To combat possible free-rider-effects, bring to light social consequences of new technology, meet EEC regulations or simply to devise new methods, evaluation studies were commissioned by federal ministries (refs. 3-4).

EVALUATING GERMAN INNOVATION POLICY

Single theme and wide scope evaluation studies have been carried out on the following topics:

- R&D incentives via tax policy (general and regional deprecation allowances on R&D investments plus R&D investment subsidies; ref. 5)
- Initial innovation programme (Erstinnovationsprogramm;ref. 6)
- State and regional business foundation grants (ref. 7)
- Experimental project on technology and innovation counselling (Modellversuch Technologie- und Innovationsberatung; refs. 8-9)
- Grants for R&D personnel (FuE-Personalkostenzuschußprogramm; refs. 10-11) and and an experimental modification of the programm (ref. 12)
- Data-processing programmes between 1967 and 1979 (ref. 13)
- Grants for contract research (Programm Auftragsforschung; ref. 14)

Further evaluation studies are currently being carried out on a micro-electronics special programme (refs. 15-16), programmes on micro-peripherals. 17-18) and the production technology programme (refs. 19-20).

A study also exists on the effects on employment of R&D promotion excepting tax measures (ref. 21), which touches on all the different research promotion measures. Studies to improve the range of promotional methods also deal with the effect of existing innovation promotion:

- Promotion of innovation demand (ref. 22)
- R&D activity in Hamburg (ref. 23)
- Innovation incentives for tradesmen (refs. 24-25)
- Promotion of R&D and innovation in SMEs by the EEC (ref. 26)
- Promotion of productional outfitting and market introduction (ref. 27)
- R&D promotion in Berlin (ref. 28).

THE "RIGHT METHOD" DISPUTE

A long standing debate as to whether quantitative or qualitative methods should be used in their studies occupied american evaluation researchers up until the early eighties. The initially prevalent quantitative approach was forced into the defensive by the appearance of qualitative methods. This gave rise to a rather fruitless style of argumentation, demanding an either/or standpoint.(ref. 29)

In a systematic comparison, Patton characterizes the two approaches in the following way. The typical qualitative approach has

- qualitative data,
- holistic-inductive design of naturalistic inquiry, and
- content or case analysis.

The traditional hypothetical deductive approach typically contains

- quantitative data from
- experimental (or quasi-experimental) research designs and
- statistical analysis. (ref. 30)

The upshot of the debate was that various elements of both approaches could and should be combined (ref. 31). Various possible combinations are shown by Patton (ref. 32).

In the case of an exploratory study on a new policy measure whose basic effects are still unknown, the following combination might, for example, be used:

- Qualitative data from pre-structured interviews in a
- natural conversational setting, interpreted by means of an
- analysis of contents, and the resulting mass of data consolidated into types by a
- cluster analysis

Theoretical demands for the use of multiple methods in evaluation research have unfortunately been more numerous than actual successfully completed studies. However, elements of the multiple methods approach now turn up in a very wide range of studies on german innovation policy.

PREVIOUS APPROACHES

The existing studies in this field cover a broad spectrum of approaches. The standard approach consists of preliminary talks with members of ministries and project sponsors, with a simple written survey sheet sent out to promoted businesses. (ref. 33) In some studies, interviews are made in the businesses involved (e.g. refs. 34) which sometimes develop into case studies (refs. 35-36). The Batelle Institute also carries out several discussion programmes with representatives of promoted firms as well as of Ministries and sponsors (ref. 37). Comprehensive inquiries also analyse the project sponsors' data or the application correspondence (ref. 38).

Prognos centres its investigations around the statistical analysis of promotion expenditure in order to determine the direct and indirect effects of research promotion on employment, and only uses consultation to supplement this. (ref. 39)

To single out the real effects of a programme, counter sample comparisons are made with check groups (refs. 40-41), or time series and trends analyses (Allesch touches on this peripherally; ref. 42). To some extent, attempts have been made to contrast a hypothetical development (i.e. without promotion measures) with the real one. (refs. 43-44)

Probably the methodically most ambitious study to date was carried out by the Fraunhofer-Institute for Systems and Innovation Research (ISI) on the R&D personnel grants programme (refs. 45-46). It was based on an idea by Bräunling et al. (ref. 47) which Meyer-Krahmer, Gielow and Kuntze developed in the course of their research. Their procedure consists in comparing innovation behaviour both in the promoted business itself before and during the programme (before-and-after comparison) and also in promoted and non-promoted SMEs carrying out R&D (counter sample comparison).

Each stage in the investigation includes a statistical analysis of the data of all businesses involved, a written survey carried out on a representative cross-section, personal consultation with a number of these plus a personal and written survey among non-promoted businesses. A relatively thorough and precise analysis of this data is carried out using a statistical programme for social scientists (SPSS), however there is no indication of any systematic analysis of the interviews.

As a whole, the methodical emphasis lies on the two written questionnaire surveys. One important result however, is the appearance of a rudimentary typology of enterprises, demonstrating the different effects of government R&D promotion.

This study is a typical example of the more ambitious multiple approach studies. A qualitative investigation is included in the initial evaluation concept and carried out in the promoted businesses. But in terms of the amount of time and methodical systematical care spent on it, it does not come up to the standard of the quantitative survey.

This placing of methodical emphasis should not be underestimated in its effect on the depth to which certain questions can be examined and, consequently, on the results of the evaluation. For example, an intensive and systematic investigation into attitudes held in the promoted businesses, e.g. the willingness to take risks, etc. is not possible, and has not yet been attempted, although there is no lack of empirical material pointing to

- a strong influence of certain attitudes on the effectivity of promotion measures; and vice versa
- a similarly great alteration of attitudes by these measures.

INFLUENCE OF ATTITUDES ON THE EFFECTIVITY OF INNOVATION POLICY

To make innovation policy work, the individual measures have to be packaged in a way that makes them acceptable to management. It is also vital to ensure that the businesses who go on to receive the aid use it for the desired purpose. These important conditions for the success of innovation policy lead evaluators to ask two questions in particular:

- Who is the clientele (i.e. beneficiaries)?
- To what extent and why are the grants really being used in the way intended by the programme's initiators?

Bearing these two problems in mind, we would like to demonstrate that many of the questions occurring in evaluation studies will need a systematic analysis of underlying managerial attitudes, in order to provide reasonable answers.

Clientele Analysis

This tells us a) who hears about the financial or advisory aid, b) accepts the idea, and c) finally receives the aid. Access and acceptance barriers (e.g. application costs procedures) often prevent the desired broad impact, and lead to a factual elimination, or at least severe under-representation of certain types and sizes of business, and geographical regions (refs. 48-49). To throw more light on this distribution problem, broad based, representative studies are necessary, together with a closer look at the causes of non-acceptance among those who should most have benefitted. We assume for the moment that non-acceptance has its roots in underlying attitudes inside business management:

- Which outlook prevents them from seeking information on government aid?
- Do they find it hard to believe that the state is suddenly interested in their well being, after years of only being interested in their taxes?
- Can the government really increase its credibility by replacing glossy brochures with personal innovation counselling?

Substitution or Free-Rider-Effects

A further good example of the advisability of using the multiple approach method to scrutinize attitudes more carefully is the substitution or free-rider-effect. This defines a situation whereby government aid is claimed without the promoted business putting the funded project into effect (refs. 50-51). The existence of this effect is undenied; however proof of its taking place, or a realistic appraisal of its extent are extraordinarily difficult to come by. Attempts have been made by Grönhaug/Fredriksen (ref. 52), Allesch et al. (ref. 53), Schinkel (ref. 54), as well as in past of a Prognos study on the effects of R&D promotion on jobs (ref. 55). In this context, it is especially worth mentioned the procedure used by Meyer-Krahmer et al. (ref. 56), as very varied, complementary methods were used, for example, counter sample comparisons with check groups. For these, obviously, very large scale surveys are necessary, which would also need to be supplemented by a scrutiny of attitudes underlying the free-rider-effect.

Such areas might be:

- How do business men experience the subsidization of their large-scale rivals or customers to the tune of millions of marks?
- How fair does ist seem to them, when they find themselves ineligible on account of formal regulations?
- How wide spread is the feeling of injustice in SMEs regarding subsidies?
- How may this feeling be increased or lessened by innovation policy?

In the same way that managerial attitudes towards innovation policy influence the effect of the policy, we must also ask if certain attitudes in business management are not similarly influenced by innovation policy.

INNOVATION POLICY AFFECTS ATTITUDES

In 1958, March and Simon voiced the opinion that innovation depended not only on the structure of a business, but also on the behavior of individuals and groups (ref. 57). In 1968, Mansfield went on to stress the role played by "non-economic variables", i.e., among other things, by attitudes:

> "A firm's financial help, as measured by its profitability, liquidity and growth rate, bears no close relationship to how long it waits before introducing a new technique.(...) The personality attributes, interest, training, and other characteristics of top and middle management may play a very important role in determining how quickly a firm introduces an innovation. The presence or absence of a few men in the right places who believe strongly in the value of the technique may make a crucial difference." (ref. 58)

Empirical studies have confirmed, for example, the influence of motivation (ref. 59) and elite values (ref. 60) on the ability of businesses to innovate. Within the discussion surrounding the subject of organizational culture, their decisive effect has been recognized:

> "Above-average product or process know-how is no longer sufficient to ensure an advantage among rivals. The critical factor is the firm's grounding, i.e. the organizational culture which carries this knowledge." (ref. 61)

By organizational culture we understand the following concept, defined by Ulrich (ref. 62), as the sum of all consciously or unconsciously cultivated and handed down symbolic or verbal stores of knowledge, together with their accompanying behavioural and thought patterns, which have developed from the experience of mastering the firm's need for internal and external stability, and which regularly find their expression in thought, word and deed by the employees of that firm.

Innovation Orientatedness as Part of Organizational Culture

We regard an orientation towards innovation as the most important part of organizational culture. Its intensity will alter: e.g. in times of unrest or insecurity it is very low, and in phases of relative stability it increases accordingly. Bleicher (ref. 63) contrasts low and high levels of innovation orientation as follows:

Stability seeker with low innovation orientation	Organization with high innovation orientation
- seeks safety in rules	- seeks spontaneous challenge
- tends towards equilibrium avoids imbalance	- initiates and uses imbalance
- averse to risk	- willing to take risks
- bureaucratic thought & action	- entrepreneurial thought & action
- tendency to conserve	- tendency to change

That innovation policy is capable of influencing orientation towards innovation has been shown by empirical studies which indicate a) that innovation promotion increases the tendency to be innovative (ref. 64) and b) that psychological barriers are overcome (ref. 65).

To the extent that research has demonstrated the importance of cultural factors for the innovation process, the question which now arises is: How far can innovation policy influence the innovation process by modifying innovation orientation?

To answer this question it is necessary to take a closer look at innovation orientation.

Analysis of Innovation Orientation

As a structural guide-line we use the Attitude Concept, and differentiate between affective, cognitive and behavioral components (ref. 66).

(i) The affective component is the emotional basis for all valuations taking place within perceptual and other cognitive processes. Here, we ask to what extent the individual has positive feelings about innovation, or if he rejects it from the start. What interests does he associate with innovation? Or is it a value in itself for him? The importance of these individual values is demonstrated in the study by Meyer-Krahmer et al. in which they describe the ingrained and unshakeable scepticism of various business managements towards R&D.

(ii) The cognitive component (ability to innovate) comprises all knowledge on matters relevant to innovation, and the competence to carry out the desired innovation. This includes relevant technical and organizational knowledge, social sensitivity and competence (e.g. ability to put things over), the degree of outward orientation, temporal horizon, but also level of argumentation regarding technical progress. In terms of the cognitive component, we can study how the individual experiences the situation and how he sees reality. Important in this context is the way ignorance relates to action. Schumpeter indicated this when he emphasized the innovation promoting effects of lack of information about the activities of the competition, as an "omnipresent threat" (ref. 67).

(iii) The behavioral component is that part of the attitude which actually affects how an individual will act in a given situation. Defined as a tendency to innovate, it comprises all forms of readiness to act in a given situation. Among these are: co-operativeness and the readiness to take risks, to patent and to utilize external R&D.

For an analysis of the effects of incentives, it is especially worth studying changes with long term consequences:

- Has the firm's management re-orientated itself? Is it more interested in technological questions? Does it have more queries on the state of new developments?
- Is a stronger general outward orientation taking place? Is external contract research or research co-operation being considered?
- What decisions have been made on personnel or their requalification? What reasons were given, and which interests did this reveal?
- Has the quality of innovation management improved? Is more emphasis being laid on intensive co-operation in the areas of R&D and marketing? Has the need for information and the informational behaviour towards new technologies become greater? Are offers of database research being accepted?

OUR OWN EXPERIENCE

We are engaged in a two year project on the influence of state innovation policy on SMEs. This project is intended to demonstrate how the effects of innovation policy on innovation orientation can best be examined. During the first seven months, we have carried out interviews in eleven businesses. In three of these we interviewed more than one person. Most of these tape-recorded interviews have already been transcribed and are being evaluated in a four person group.

Data Collection

(i) Selection of businesses. Our aim is to achieve the completest possible picture of all conceivable effects of innovation policy on the tendency towards innovation. Following Patton's example (ref. 68), we decided to select a wide variety of businesses: new and old (from 3 to 225 years standing), from varying sectors (bio-technology, electronics, mechanical engineering, optical instruments, foundry), and from varying regions (from village to city). We also set out to survey firms where conflicts about innovation subsidies had arisen between the Works Council and management. But even with the help of the Trades Unions, we were unable to find a business in which such problems had arisen.

(ii) Carrying out the Interviews. The interviews touch on a number of subjects which necessitate an atmosphere of complete confidentiality. It isn't simply a matter of how the firm organized the innovation process, or of where they get their technical know-how or sales information from. More important is what they actually did with the subsidies received. Were the funds needed for and spent on the project applied for? Would a subsidized investment have been carried out at that point anyway? Did the funds simply speed things up? Which problems were talked about during advisory consultations?

(iii) Visiting the Business. A visit to the firm gives many clues as to the prevailing atmosphere there. Is it friendly or reserved? How do the employees react to their boss or to the unfamiliar visitor? How is the working area furnished? How open or secluded are the various working areas from each other? How up to date is their equipment (machines, computers, telephone systems)? These provide many indicators of readiness to innovate which would hardly come to light in a questionnaire or even a conversation. We took notes of all our visits to aid in the interpretation of the interviews.

Data Analysis

(i) Transcription. We make this as exact as possible. Pauses, laughter, ahems etc. are included. Every transcribed interview is then proof-read to correct mistakes, making this a relatively time-consuming part of the procedure.

(ii) Interpretation. Each evaluator first reads the interview and notes his or her impression of the interview as a whole. Questions to be borne in mind are:

- How do the people in the business see themselves?
- How great is the receptiveness towards innovation?
- How does the business handle its subsidies?

This creates a subjective holistic picture of the prevailing attitudes in the business.

Following this, the group discusses which similarities or differences have cropped up in their first impressions and what these consist of in particular. Passages drawn on in the text to support impressions are examined to see if they justify these.

With this method of text interpretation, the initial selection of project members is of central importance. We made sure that they would be able to work together to take up and clarify conflicts. In the evaluation process, it is vital for people to be committed to voicing and justifying opinions, but also to be capable of altering them after discussion. The more open the relationships inside the evaluation group are, the sooner new insights are achieved.

(iii) Coding. One problem of our method is the handling and coding of the great mass of interview material. We use a personal computer and store the interviews in a database. This helps to preserve anonymity, and allows flexible coding. We code according to a feature list, which ideally includes all features deemed relevant during interpretation. We aim to constantly redevelop this list, which serves as a mirror of the evaluation process.

CONCLUSION

Attitudes held by promoted businesses are of double importance for innovation policy. On the one hand they determine the way firms treat innovation subsidies (acceptance of measures / free-rider-effect), and on the other, empirical studies suggest that they can themselves be modified by innovation policy.

These attitudes cannot be assessed by purely quantitative methods, however. Even a multiple approach method only becomes viable if sufficient attention is paid to the specific demands of the qualitative section at the planning and funds-allocation stage. Our experiences in carrying out an evaluation study with qualitative methods point to the need for good initial contacts to the promoted businesses, and experience in interpretative procedures. We hope that if government sponsors can bring themselves to dispense with the idea that truth lies only in quantifyable facts and figures, and recognise the feasibility of surveying attitudes, they will see the need for allocating funds for - apparently expensive - evaluation programmes. In doing so, they could benefit from a long-standing methodical discussion and pave the way for fundamental new impulses in innovation policy.

ACKNOWLEDGEMENT

Support for the research underlying this paper was provided by the Deutsche Forschungsgemeinschaft (ref. no. Sta 162-5). My thanks are due to Professor Dr. W. H. Staehle (project leader), Thomas Gelshorn, Claudia von Rostowsky and Claudia Scholle for their help and cooperation.

REFERENCES

1 Bundesministerium für Forschung und Technologie (Ed.), Bundesbericht Forschung 1984, Bundestagsdrucksache 10/1543, Dr. Hans Heger, Bonn 1984

2 Bundesministerium für Forschung und Technologie (Ed.), Ratgeber Forschung und Technologie, Fördermöglichkeiten und Beratungshilfen, Köln 1986

3 H. Kohn, Der Einsatz von Wirkungsanalysen im Tätigkeitsfeld des Bundesministeriums für Wirtschaft, in: G.-M. Hellstern and H. Wollmann (Eds.), Handbuch zur Evaluierungsforschung, Vol. 1, Schriften des Zentralinstituts für Sozialwissenschaftliche Forschung der Freien Universität Berlin 35, Opladen, 1984 pp. 320-325

4 M. Széblábi, Erfolgsbewertung in der Forschungsförderung, Probleme und Verfahren im Bundesministerium für Forschung und Technologie, in: G.-M. Hellstern and H. Wollmann (Eds.), Handbuch zur Evaluierungsforschung, Vol. 1 Schriften des Zentralinstituts für Sozialwissenschaftliche Forschung der Freien Universität Berlin 35, Opladen, 1984, pp. 345-358

5 K. C. Röthlingshöfer and U. Sprenger, Effizienz der indirekten steuerlichen Forschungsförderung, Berlin, München, 1977

6 B. Hoffmann, Das Erstinnovationsprogramm als staatliche Maßnahme zur Förderung industrieller Innovationen, Bonn, 1978

7 S. Schinkel, Auswirkungen der Existenzgründungsprogramme, Erfahrungen der Betriebe in der Praxis (Beiträge zur Mittelstandsforschung, ed. by Institut für Mittelstandsforschung, Forschungsgruppe Köln, Vol. 59), Göttingen, 1980

8 G. Bräunling, G. Jaeckel, J. Pinter, H. R. Schöler, G. H. Walter, Darstellung, Bewertung und Perspektiven öffentlich rechtlich geförderter Pilotvorhaben zur Innovationsberatung - eine Zwischenbilanz, ISI-Fraunhofer-Institut für Systemtechnik und Innovationsforschung, ISI-P-7/81, Karlsruhe 1981

9 G. Bräunling, Publicly funded organizations assisting innovation and technology transfer in small and medium sized companies, in: TII - Europäische Vereinigung für den Transfer von Technologien, Innovation und industrieller Information (Ed.), Proc. Int. Seminar: How to help SMEs to acquire and to integrate new technologies, Frankfurt (Main), March 6-7, 1985, Luxembourg 1985

10 F. Meyer-Krahmer, G. Gielow and U. Kuntze, Innovationsförderung bei kleinen und mittleren Unternehmen, Frankfurt (Main), New York, 1982

11 F. Meyer-Krahmer, G. Gielow and U. Kuntze, Wirkungsanalyse der Zuschüsse für Personal in Forschung und Entwicklung - Endbericht an den Bundesminister für Wirtschaft, Bonn, Forschungsbericht des ISI-Fraunhofer-Institut für Systemtechnik und Innovationsforschung, ISI-B-4-84, Karlsruhe, 1984

12 K. Brockhoff, The measurement of goal attainment of governmental R & D support, Research Policy 12, 1983, 171-182

13 Arthur D. Little, SRI International, Die Entwicklung der Datenverarbeitung in der Bundesrepublik Deutschland - Programmbewertung der DV-Förderung des BMFT 1967 bis 1979, Wiesbaden 1982

14 J. Allesch, H. Fiedler and H.-J. Martin, Förderung der Auftragsforschung, Zwischenbilanz der Fördermaßnahme "Externe Vertragsforschung" - eine Untersuchung der TU Berlin (Technologie-Transfer-Stelle) in Zusammenarbeit mit ISI Karlsruhe im Auftrag des BMFT, Köln 1983

15 VDI-Technologiezentrum, Sonderprogramm "Anwendung der Mikroelektronik" des Bundesministers für Forschung und Technologie, erster Erfahrungsbericht, Berlin, 1983

16 VDI-Technologiezentrum, Sonderprogramm "Anwendung der Mikroelektronik" des Bundesministers für Forschung und Technologie, zweiter Erfahrungsbericht, Berlin, 1985
17 VDI-Technologiezentrum, Förderungsschwerpunkt "Mikroperipherik-Verbundvorhaben", erster Erfahrungsbericht, Berlin, 1986
18 VDI/VDE-Technologiezentrum Informationstechnik GmbH, Förderungsschwerpunkt "Mikroperipherik", zweiter Erfahrungsbericht, Berlin, 1987
19 R. Brüning, R. Friedrich and V. Korndörfer, Wirkungsanalyse der indirekt-spezifischen Förderung zur Entwicklung von Industrierobotern, Handhabungssystemen sowie zugehöriger intelligenter Peripherie, erste ausgewählte Ergebnisse, in: I. Bey and H. Mense (Ed.), Proc. Seminar on Bewertung von Entwicklung und Einsatz moderner Fertigungstechnologien, Karlsruhe, March 12, 1986, Kernforschungszentrum Karlsruhe, 1986
20 G. Lay, M. Boffo, J. Fix-Sterz, U. Kuntze and F. Meyer-Krahmer, Erste Ergebnisse der Wirkungsanalyse der indirekt-spezifischen Förderung zur betrieblichen Anwendung von CADCAM-Systemen, in: I. Bey and H. Mense (Eds.), Proc. Seminar on Bewertung von Entwicklung und Einsatz moderner Fertigungstechnologien, Karlsruhe, March 12, 1986, Forschungsbericht KfK-PFT 119, Kernforschungszentrum Karlsruhe, 1986
21 BMFT - Bundesminister für Forschung und Technologie (Ed.), Arbeitsplatzwirkungen der Forschungsförderung - Versuch einer quantitativen Abschätzung, Technologie und Beschäftigung, Vol. 5, Econ, Düsseldorf, Wien, 1980
22 H. Geschka and U. Alter, Industrieeigene Vorausschau und Bewertung zukünftiger Entwicklungen und deren Einfluß auf die Forschungs- und Innovationspraxis ausgewählter Industriebranchen - Kooperationsformen zwischen Staat und Wirtschaft. Bericht des Batelle-Instituts für den BMFT, Frankfurt, 1976
23 H. Körner, Forschung und Entwicklung in der Hamburger Wirtschaft. Auswertung einer Befragung im verarbeitenden Gewerbe unter besonderer Berücksichtigung kleiner und mittlerer Betriebe. Untersuchung im Auftrag der Behörde für Wirtschaft, Verkehr und Landwirtschaft in Zusammenarbeit mit dem Forschungsrat der Freien und Hansestadt Hamburg. Hamburg, 1978
24 W. Koschorke and G. Schilling, Strategien einer verstärkten Förderung von Entwicklungsarbeiten im Handwerk - Schlußbericht, Bonn, Hannover, 1979
25 G. Schilling and L. Kreft, Innovationsförderung im Handwerk (AGR - Arbeitsgemeinschaft für Rationalisierung des Landes Nordrhein-Westfalen, ed. by Minister für Wissenschaft und Forschung, Vol. 209, Dortmund, 1980
26 K. C. Röthlingshöfer, Möglichkeiten der Förderung von Forschung, Entwicklung und Innovation in kleinen und mittleren Industrieunternehmen durch die europäischen Gemeinschaften, Ifo-Institut für Wirtschaftsforschung, München 1980
27 L. Rall and R. Pfeiffer, Die Förderung der Produktionseinrichtung und Markteinführung bei kleinen und mittleren Unternehmen in ordnungspolitischer Sicht - Gutachten im Auftrag des Bundesministers für Wirtschaft, ed. by Institut für Angewandte Wirtschaftsforschung Tübingen, Forschungsberichte, Serie A, Nr. 28, Tübingen, 1981
28 K. Hornschild, Forschung und Entwicklung im verarbeitenden Gewerbe von Berlin (West) - eine Repräsentativerhebung für die Jahre 1977-1981, DIW-Beiträge zur Strukturforschung 76, Berlin, 1983
29 R. F. Conner, D. G. Altman and C. Jackson, 1984: A Brave New World for Evaluation? in: R. F. Conner, D. G. Altman and C. Jackson (Eds.), Evaluation Studies, Review Annual, Vol. 9, Beverly Hills, London, New Delhi, 1984
30 M. Q. Patton, Utilization-Focused Evaluation, Beverly Hills, London, 1978, pp. 199-238
31 C. S. Reichardt and Th. D. Cook, Beyond Qualitative Versus Quantitative Methods, in: Th. D. Cook and C. S. Reichardt, Qualitative and Quantitative Methods in Evaluation Research, Beverly Hills, London, 1979
32 M. Q. Patton, Qualitative Evaluation Methods, Beverly Hills, London, 1980, pp. 108-117
33 Examples are B. Hoffmann, Das Erstinnovationsprogramm... op. cit.;
K. Hornschild, Forschung und Entwicklung im verarbeitenden... op. cit.;
H. Körner, Forschung und entwicklung in der Hamburger Wirtschaft... op. cit.

34 W. Koschorke and G. Schilling, Strategien einer verstärkten ... op. cit. and L. Rall and R. Pfeiffer, Die Förderung ... op. cit.
35 K. C. Röthlingshöfer et al., Effizienz der... op. cit.
36 T. Ellwein and W. Bruder, Innovationsorientierte Regionalpolitik, (Beiträge zur sozialwissenschaftlichen Forschung, Vol. 31), Westdeutscher Verlag, 1982
37 H. Geschka et al., Industrieeigene Vorausschau... op. cit.
38 e.g. J. Allesch et al., Förderung der Auftragsforschung... op. cit.
39 BMFT (Ed.), Arbeitsplatzwirkungen der Forschungsförderung... op. cit
40 T. Ellwein et al., Innovationsorientierte... op. cit.
41 J. Allesch et al., Förderung der Auftragsforschung... op. cit.
42 ibid.
43 K. Grönhaug and T. Fredriksen, Governmental innovation support in Norway - Micro- and macro-level effects, Research Policy 13 (1984) 165-173
44 S. Schinkel, Auswirkungen der Existenzgründungsprogramme... op. cit.
45 F. Meyer-Krahmer et al., Innovationsförderung... op. cit.
46 F. Meyer-Krahmer et al., Wirkungsanalyse... op. cit.
47 G. Bräunling, P. Georgieff, U. Kuntze, H. Legler, F. Meyer-Krahmer, Zur Konzeption und Organisation von Wirkungsanalysen in der Technologiepolitik, ISI-Fraunhofer-Institut für Systemtechnik und Innovationsforschung, ISI-B10-79, Karlsruhe 1979
48 K. C. Röthlingshöfer et al., Effizienz..., op. cit. pp. 68-69
49 F. Meyer-Krahmer et al., Wirkungsanalyse..., pp. 127-133
50 T. J. Allen, J. M. Utterback, M. A. Sirbu, N. A. Ashford and J. H. Hollomon, Government influence on the process of innovation in Europe and Japan, Research Policy 7 (1978) 124-149
51 M. A. Sirbu, Government Aid for the Development of Innovative Technology: Lessons from the French, Research Policy 7 (1978)
52 K. Grönhaug et al., Governmental innovation... op. cit.
53 J. Allesch et al., Förderung der Auftragsforschung... op. cit.
54 S. Schinkel, Auswirkungen... op.cit.
55 BMFT, Arbeitsplatzwirkungen... op. cit.
56 F. Meyer-Krahmer et al., Wirkungsanalyse... op. cit.
57 J. G. March and H. A. Simon, Organizations, Wiley, New York, 1958, p. 197
58 E. Mansfield, Industrial Research and Technological Innovation, an Economic Analysis, New York, 1968, p. 204
59 L. B. Mohr, Determinants of Innovation in Organization, in: The American Political Science Review, Vol. 63 (March 1969) 111-126
60 J. Hage and R. Dewar, Elite Values versus Organizational Structure in Predicting Innovation, in: AScQ 18 (1973) 279-290
61 C. Pümpin, J.-M. Kobi and H. A. Wüthrich, Unternehmungskultur, Basis strategischer Profilierung erfolgreicher Unternehmen, in: Die Orientierung, (Schriftenreihe der Schweizerischen Volksbank, Vol. 35), Bern, 1985, p. 7 (our translation)
62 P. Ulrich, Systemsteuerung und Kulturentwicklung, Die Unternehmung 38 (4) (1984) 303-325
63 K. Bleicher, Strukturen und Kulturen der Organisation im Umbruch: Herausforderung für den Organisator, zfo (2) (1986) 97-106
64 G. Mensch, Staatliche und unternehmerische Instrumente der Innovationsförderung, dp/77-45, Wissenschaftszentrum Berlin, 1977, S. 13
65 K. Grönhaug et al., Governmental innovation... op. cit., p.170
66 M. J. Rosenberg and C. I. Hovland, Cognitive, Affective, and Behavioral Components of Attitudes in: M. J. Rosenberg, C. I. Hovland, W. J. McGuire, R. P. Abelson, J. W. Brehm (Eds.), Attitude Organisation and Change, Yale University Press, New Haven and London, 1960, pp. 1-14
67 J. A. Schumpeter, Kapitalismus, Sozialismus und Demokratie, 5th edn., München 1980
68 M. Q. Patton, Qualitative Evaluation Methods, ... op. cit., p. 102

Innovation: Adaptation and Growth, edited by R. Rothwell and J. Bessant
Elsevier Science Publishers B.V., Amsterdam, 1987 — Printed in The Netherlands

EXTERNAL CORPORATE VENTURING: THE EXPERIENCE TO DATE

P. G. OAKLEY

Department of Business Management, Brighton Business School, Brighton Polytechnic, Mithras House, Lewes Road, Brighton, East Sussex, BN2 4AT (England)

SUMMARY

External corporate venturing holds considerable promise as a larger company approach to the development of new businesses. But despite the biggest European venture capital industry, U.K. corporate involvement in the organised venture capital industry is very limited, although in its widest sense external venturing is probably more prevalent than commonly believed. It is shown that external venturing has met with greater success than internal venturing, and is most suited for accessing new technology for new markets. The viability of venturing can be improved when strongly supported by the corporate board, under taken with long time horizons and managed at arms length. At a first approach to external venturing, a creeping commitment is the preferred strategy, starting with the lowest risk route of investment in an independent venture capital fund.

INTRODUCTION

At a time of rapid change large corporations are under pressure to innovate quickly - especially when their core activities are mature and growing slowly. But bureaucratic structures often stifle new business development. Hence, in an endeavour to emulate the more successful innovative record of smaller firms (ref. 1) many established corporations have adopted corporate venturing as a means of gaining early access to new technologies whilst giving their own organisation a much needed dash of entrepreneurial vigour.

Janus like, this approach has two faces - one internal, the other external although hybrid variants are possible -

FIGURE 1

The Range of Corporate Venturing Activities

EXTERNAL				INTERNAL
Venture Capital	Venture Nurturing	Venture Spin-off	Joint Venture	Intrapreneuring

——————— Increasing risk and corporate commitment ⟶

Source: After Roberts & Pinchot (refs. 2, 3)

Internal venturing was first popularised by 3M and DuPont in the late 1960's and its use has waxed and waned ever since. However, although it can be made to work, as witnessed by the successes of 3M, Hewlett-Packard and Monsanto, experience has generally been disappointing (refs. 4-6). This is not perhaps surprising. Roberts (ref. 2) speculated that internal venturing is the riskiest of the possible forms of venture management as it necessitates the highest level of corporate commitment. At the other end of the scale the provision of venture capital was seen as the least risky strategy since it could involve a lower level of corporate exposure and participation. In consequence it might be expected that external venturing would be widely employed. But relative to the U.S.A. the practice does not appear to be so widespread in the U.K. (ref. 7). This paper therefore focuses upon U.K. experience of a promising but 'under-utilised' approach to new business development. In doing so U.S., European and internal venture studies are cited to provide illustrations and performance 'benchmarks'.

U.K. EXTERNAL CORPORATE VENTURING

Firstly it should be noted that in this context external corporate venturing implies the corporate venturer is involved in taking a minority equity stake in a young growth business. This approach is based upon the premise that small companies tend to react to market changes and pick up new technologies faster than big ones. Thus corporate investors in them are better able to latch onto new ideas before competitors. Whilst it is difficult to gauge the full extent of the practice, data is available on corporate investment in independent venture capital funds, which then go on to reinvest this and other monies. Before 1984, according to Venture Economics (ref. 8), no U.K. corporation undertook this approach, but in 1984 a modest start was made and £17m. was committed. Between 1984-86 (inclusive) the average annual figure was £13m., 5% of an average annual total of about £280m. raised by independent venture capital funds. However, only 4 U.K. venture capital groups are known to have raised capital from industrial corporations. In contrast, between 1978-84 (inclusive) U.S. corporations committed an average annual sum of $198m., 14% of an average annual total of approximately $1.4bn. raised by independent funds (ref. 9), and over 100 American corporations are involved in this method of corporate venturing. By way of partial explanation, the organised U.K. venture capital industry, in its present form, only dates from about 1980, whilst the U.S. industry dates back to the early post war years. A more venturesome U.S. corporate climate is therefore to be expected. Further, the figures cited above only relate to corporate investment in managed funds. A range of possibilities are available and these are outlined in Figure 2 below -

FIGURE 2

The Range of External Corporate Venturing Possibilities

Investment through:-	Independent venture capital funds	Internally managed venture capital funds	One-off's e.g. Corporate partner-ing	Spin-outs from parent - often with v.c. partner
Venture type:-	Venture Capital/Nurturing			Spin-outs

Combining two studies by Littler and Sweeting (ref. 10) incorporating a total of 101 selected U.K. companies, 42% had experimented with variants of venture capital/nurturing, whilst 9% had experimented with spin-out venturing. It is therefore probable that external corporate venturing is more common than often reported (refs. 7, 8, 11 and 12) although less frequently employed than in the U.S.A.

Discussion and Examples of Venturing

The various methods of external corporate venturing shown in Figure 2 are now explored in more detail.

The most publicised approach is investment through independent funds - a corporation typically investing in 2 to 5 separate funds at a total cost of from £2m. to £5m. (ref. 12). In this way the firm gains a spread of ventures and several avenues to useful technological developments. The classic example of how fund investment can help a corporation innovate is from Monsanto the U.S. chemicals group which, since the early 1970's, has put about \$50m. into 9 funds around the world, mostly managed by the U.K. venture group Advent. Monsanto believes that this activity accelerated its move into biotechnology by two years.

Other corporations known to have invested in U.K. independent venture funds are Air Products U.K., American Hospital Supplies, Baker International, Elf Aquitaine, Molson, Johnson & Johnson, Siemens, B.P., G.K.N. and Pilkingtons - but of these, only the last three are of U.K. parentage.

A variant of the fund investment theme is to launch an internally managed venture capital fund, though these tend to work best when given plenty of independence. Again, Monsanto has tried this method, as have Exxon and General Electric. It generally involves higher capital outlays, and some U.S. corporations committed up to \$100m. a year to their investment programmes (ref. 12). From the U.K.,Pilkingtons, Ferranti and Thorn EMI have adopted this approach. Of these only Pilkingtons has invested 'widely' in the U.K., but its unfocused strategy based upon the differing aims of diversification and regenerating the economy of the north west has performed 'poorly'. From 10 investments, representing £2.5m., only 2 are doing well, a result attributed in part to the 'half time' commitment of the fund managers who double up as divisional heads. Alternatively, Ferranti and Thorn EMI provide examples of a focused approach. Both have U.S. based venture funds - providing them with a window on high tech developments as well as helping to identify potentially useful products for their European marketing networks. Management is full-time whilst investments have invariably been alongside experienced U.S. venture capital firms.

More recently corporate venturers have tended to concentrate on picking their own one-off investments rather than going through funds. This practice sometimes called corporate venture partnering or direct strategic investment seeks to establish a mutually beneficial alliance in which the junior partner enjoys not just financial support but also managerial and/or marketing help.

In the U.S. the number of such alliances in venture backed companies has increased from 30 in 1980 to 245 in 1985 (ref. 13). This is partly due to the sophistication of the U.S. venture capital industry where news of promising young companies travels faster and further than in the relatively confused and younger European venture capital market. But some European companies have always preferred this route. Between 1979 and 1986 Olivetti acquired stakes in 25 U.S. companies as well as investing in Sphinx, a U.K. supplier of Unix based software. Ultramar , Cambridge Electronics Industries and Ferranti have also made such strategic investments, all alongside U.K. venture capital funds.

Large corporations adopt an even more parental role in the venture based spin-out, a way of commercialising otherwise redundant or peripheral research ideas. An example of this approach is Rank Xerox's 25% stake in 'Artificial Intelligence', a £2.6m. turnover company that sells 70% of its turnover to the Group (ref. 14). In 1983 Rank Xerox was too busy launching new products to take up an artificial intelligence proposal. By allowing the originator to go independent Rank Xerox avoided having to stifle the opportunity, and by giving the latent entrepreneur a controlling stake, the business developed far faster than could have been possible within the organisation.

Rank Xerox is among a small group of European firms to have employed spin-outs - using the example of such corporate U.S. giants as IBM, GM and AT&T - albeit Rank Xerox originally showing greater concern for saving head office costs. A more deliberate approach was adopted at the outset by Ferranti and Siemens, with Siemens offering a model incremental approach to the spin-out. The new company only leaves the corporate fold by gradual degrees so as to soften the difficult transition from employee to proprietor. For about 2 years the 'candidate' entrepreneur remains an employee until the venture is good enough to raise outside capital on commercial terms. At this stage the employee becomes self-employed, the only condition being that Siemens has an option over 20-25% of the equity.

Other firms such as Kodak UK, Tetrapak and ICI's agricultural division have taken a slightly different tack by using spin-outs to develop possible diversification opportunities. But ICI's approach indicates how an over protective policy can hinder rather than help spin-outs. Of 9 ventures since 1984 all but 2 are running behind budget (ref. 11) - possibly because, unlike Siemens and Rank Xerox, they get access to soft loans and their managers remain salaried ICI employees with no equity interest in their enterprise. ICI's response to under performance has been to get 'tougher', but it remains to be seen whether more 'stick' can succeed without an equity 'carrot'.

Interestingly a recent attempt by 3i (Investors in Industry) to enter into venture partnerships with large corporations only found two smaller partners for whom the spin-out represented an attractive 'off balance sheet' method for launching new products/expanding geographically without having to set up a subsidiary. As Geoffrey Taylor of 3i Ventures remarked "It just shows the cultural barriers we are up against .... in big companies" (ref. 14).

EXTERNAL V. INTERNAL CORPORATE VENTURING

Whilst not exhaustive, the corporate examples given of the various external venture possibilities include many of the known U.K. participants. It is therefore appropriate to ask whether the reluctance of U.K. corporations to engage in this form of new business development is justified? To answer this question the performance of internal and external ventures are compared, although because of the recency and limited nature of U.K. experience the studies cited are American.

Generally internal ventures have not fared well. Evidence from the Profit Impact of Marketing Strategy (P.I.M.S.) data base shows that up to 80% of internal corporate ventures collapse (ref. 15), and Wesley Morse (ref. 5) has recently highlighted the growing dissatisfaction within corporations over 'intrapreneurship'. Weiss (ref. 16), comparing independent start-ups with internal corporate ventures (derived from Biggadike's P.I.M.S. analysis - ref. 17) showed that independent start-ups did significantly better than internal ventures. Since external venturing is very much concerned with capturing the benefits associated with small independent firms this result indicates the relative advantages to be derived from external venturing. Support for this thesis is also provided by other P.I.M.S. data which reveals that on average internal corporate ventures do not show profits for 7 years whilst businesses funded by venture capitalists are profitable after only 3 years (ref. 18). However, it should be noted that venture capitalists counsel against expectations of early financial returns.

Exxon provides an interesting example of the benefits of external venturing. Between 1970 and 1980, 19 internal and 18 external ventures were funded by the Exxon Enterprises Division (ref. 19). None of the internal ventures was profitable. In contrast, 6 of the external ventures were acquired by Exxon and the remaining 12 made a $218m. return on a total investment of $12m.

The evidence therefore suggests that external venturing is more successful than internal venturing. Corroboration also comes from a study of the relative risks of either funding technological innovation through venture capital funds or through separate one-off ventures (ref. 20). Using equity beta values as a measure of risk, quoted high tech venture capital firms were found to be no riskier than the average for all companies. In contrast, single R.&D. intensive firms were much riskier. This result is not in itself surprising, but pulling together the research evidence it is clear that Roberts' 1980 advice is probably correct (ref. 2). External venturing is more effective than internal venturing whilst portfolio methods are less risky than one-off investments.

CHOOSING AND IMPLEMENTING NEW BUSINESS ENTRY STRATEGIES

In the previous section it was argued that external venturing has much to commend it. However this is but one of several methods for entering new businesses and the appropriate choice amongst alternative strategies is a key management issue. This section seeks to clarify the conditions under which external venturing is appropriate.

Technology and Market Considerations

Together with corporate venturing, acquisition, licensing and joint ventures form the range of business development possibilities. They may be arranged into a matrix according to the degree of corporate 'familiarity' with relevant markets and technologies (ref. 21).

FIGURE 3

Familiarity Matrix and Preferred Entry Mechanisms

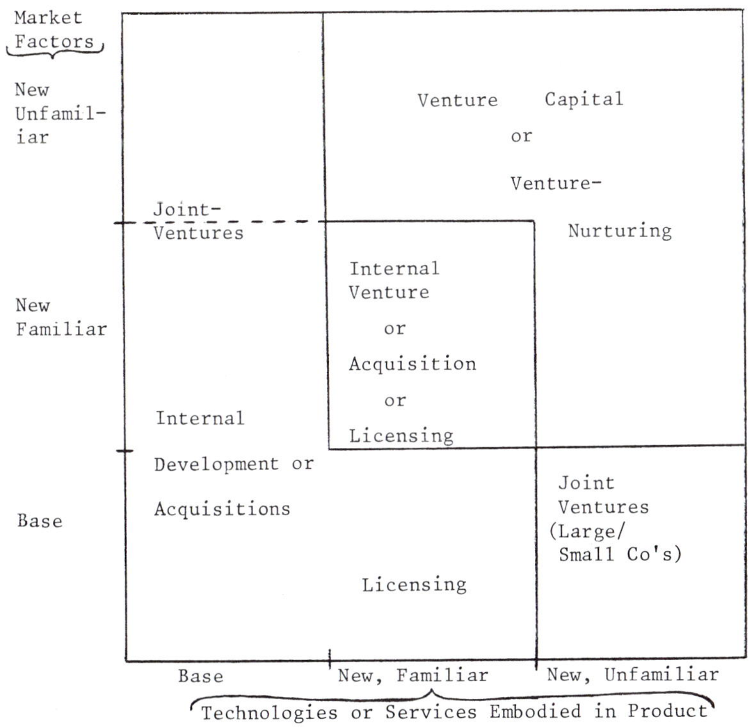

Source: After Roberts and Berry (ref. 21)

Referring to Figure 3, it is proposed that venturing is most appropriate where both markets and technologies are new and unfamiliar. Internal venturing is recommended where both markets and technologies are new BUT familiar. These recommendations also find a measure of support from an earlier analysis by Rothwell (ref. 22).

Testing the matrix against an anonymous corporation's 1971-77 experiences (ref. 21) provided evidence for its validity. All high corporate involvement mechanisms in familiar sectors were successful. But in unfamiliar areas, high involvement did not work. On the contrary, success in unfamiliar sectors seemed to be related to limiting the corporate input into the decision making process to low levels until familiarity with the new area had developed.

Monsanto has followed this approach towards its interest in biotechnology. Their first investment was through a venture capital fund. Interaction with investees was close - organised through in-house seminars - but direct management involvement was minimal. Only as internal familiarity built up did Monsanto decide to commit substantial resources to internal ventures.

Managerial Requirements for Successful External Venturing

The first point to be made is that there is no one recipe for success. Requirements inevitably change according to corporate purpose and culture. Secondly, since external venturing covers a range of approaches, what is appropriate for investment in a venture capital fund can differ from a one-off investment in a spin-out. However, the literature reveals several common ingredients for success.

(i) Senior Management Commitment. Chief executive officer (C.E.O.) commitment is probably the most important requirement (ref. 23). But, as Tony Lorenz of the venture capital firm Equity Capital for Industry points out, general board support is also important (ref. 24), 'Splits over corporate venturing usually develop on main boards between technical and financial directors, with the money men against'. He thinks this may help explain why in the U.S. - where marketing men do better in getting top jobs - companies are more ready to take the risks involved and wait for their returns.

(ii) Realistic Goals and Expectations. For large corporations venture capital is not a valid approach to diversification and overall growth, although it can provide attractive financial returns (refs. 25, 23 and 19). Rather, as discussed earlier, it is a means of accessing new technology, often in new markets. Consequently, corporate ventures require long term persistence (ref. 2) preferably on a narrowly defined front (ref. 11).

(iii) Arms Length Organisational Arrangements. Venturing responsibility is best vested in a (semi) autonomous group, recruited for their entrepreneurial flair (e.g. from the venture capital community). Reporting should be direct to the C.E.O. (ref. 23) with performance criteria reflecting the longer time horizon of new ventures. A portfolio is unlikely to break-even before the passing of 5 or 6 years. In this period corporate involvement should be minimised, thereby increasing venture management's motivation and concern about profitability, cost control, focused product development and competitive realities (ref. 19). From this it follows that large companies will have a better chance of maintaining a sense of creativity and commitment if they only take a minority stake. Whilst this may run counter to corporate instincts, tight control is anathema to success.

One issue not addressed in the venture literature, although of general concern, is that of the 'not invented here' syndrome. Since the key role of external venturing is the more effective acquisition of new technology it is clearly important to ensure that in-house R.&D. personnel are fully informed of external developments.

CONCLUSIONS

This analysis has shown that external corporate venturing can be an important component in the larger corporations' new business development programme. Its particular virtue is that as a technique it is more likely to 'capture' the entrepreneurial spirit, often so important in developing innovative new products and services. Further, as an alternative to internal venturing the available data indicates that to date external venturing has been more successful than intrapreneuring. But from the spectrum of external venturing possibilities it is important that senior management clearly define aims and methods. External venturing seems to be most successful where both technology and markets are new and when responsibility for venturing is vested in autonomous, entrepreneurial units.

Amongst the range of approaches, choice will clearly depend upon corporate circumstances. However, where there is no previous experience a form of creeping commitment seems most appropriate. Initially, investment in an independent fund is probably the wisest course. This method has the merit of securing both experienced venture managers and a spread of investees. This reduces the risks for a modest outlay. The number of UK independent funds, at about 55, are also now sufficient to ensure that a satisfactory match between corporate intentions and fund orientation can be achieved. With increasing experience the corporation may extend its activities into 'spin-outs' and 'one-offs', possibly co-investing with an independent fund.

REFERENCES

1 K. Pavitt, M. Robson and J. Townsend, The Size Distribution of Innovating Firms in the U.K. 1945-83, The Journal of Industrial Economics, March (1987) 297-315.

2 E. B. Roberts, New Ventures for Corporate Growth, Harvard Business Review, July-August (1980) 134-142.

3 G. Pinchot, Intrapreneuring, Harper & Row, New York, 1985.

4 D. A. Littler and R. C. Sweeting, Business Innovation in the U.K., R&D Management, 14(1) (1984) 1-9.

5 C. Wesley Morse, The Delusion of Intrapreneurship, Long Range Planning, 19(6) (1986) 92-95.

6 N. D. Fast, The Future of Industrial New Venture Departments, Industrial Marketing Management, 8 (1979) 264-273.

7 Special Report: Current Issues Facing the U.K. Venture Capital Industry, U.K. Venture Capital Journal, March (1987) 10-17.

8 Special Report: Review of 1985, U.K. Venture Capital Journal, May (1986) 8-19.

9 Business Brief: American Venture Capital Grows Up, The Economist, Nov. 23 (1985) 86-87.

10 D. A. Littler and R. C. Sweeting, Corporate Development - Preferred Strategies in U.K. Companies, 20(2) (1987) 125-131.

11 W. Dawkins, Why Windows on Technology are Difficult to Open, Financial Times, June 24 (1986) 20.

12 Special Report: Major Corporations and Venture Capital, U.K. Venture Capital Journal, Jan. (1985) 10-14.

13 Corporate Partnering Ventures into the '80s, U.K. Venture Capital Journal, Jan. (1986) 31-33.
14 W. Dawkins, How Peripheral Ideas Can Take Centre Stage, Financial Times, July 1 (1986) 21.
15 J. Guiniven, Ventures: Pitfalls and Pay-Offs, Management Today, August (1986) 49-51.
16 L. A. Weiss, Start-up Businesses: A Comparison of Performances, Sloan Management Review, Fall (1981) 37-53.
17 R. Biggadike, The Risky Business of Diversification, Harvard Business Review, May-June (1979) 103-111.
18 G. De Souza, The Best Strategies for Corporate Venturing, Planning Review, March (1986) 12-14.
19 H. B. Sykes, Lessons From a New Ventures Program, Harvard Business Review, May-June (1986) 69-74.
20 M. Statman and T. T. Tyebjee, The Risk of Investment in Technological Innovation, I.E.E.E. Transactions in Engineering Management, E.M.31 (4) Nov. (1984) 165-171.
21 E. B. Roberts and C. A. Berry, Entering New Businesses: Selecting Strategies for Success, Sloan Management Review, Spring (1985) 3-17.
22 R. Rothwell, From Invention to New Business via the New Venture Approach, Management Decision, 13(1) (1975) 11-21.
23 R. P. Greenthal and J. A. Larson, Venturing Into Venture Capital, The McKinsey Quarterly, Spring (1983) 70-79.
24 I. Hamilton-Fazey, Much Thought Little Action, Financial Times, Dec. 3 (1985) VI.
25 G. F. Hardymon et al, When Corporate Venture Capital Doesn't Work, Harvard Business Review, May-June (1983) 114-120.

Innovation: Adaptation and Growth, edited by R. Rothwell and J. Bessant
Elsevier Science Publishers B.V., Amsterdam, 1987 — Printed in The Netherlands

THE IMPLICATIONS OF INTEGRATED TECHNOLOGY FOR MANUFACTURING ENGINEERING STRATEGY

B W Haywood and J R Bessant
Innovation Research Group, Brighton Business School, Brighton Polytechnic, Moulsecoomb, Brighton (United Kingdom)

ABSTRACT

Considerable interest is currently being expressed in the concept of Computer Integrated Manufacturing (CIM). In its broadest sense, CIM refers to the integration of all aspects of manufacturing within a networked computer system, offering the possibility of controlling and optimising the performance of a factory as if it were a single machine.

However, a point made in our earlier work (1) - and echoed in other studies - is that many of the benefits of using technologies such as Flexible Manufacturing Systems (FMS) derive, not so much from the technology itself, but rather from the way in which it is used and the organisational changes it brings.

Internal and External Pressures on Companies

Competitive internal pressures within firms, and particularly those associated with small to medium batch manufacture, include a wide range of problems, such as:

- long production lead times
- high inventory levels
- low machine utilisation
- queuing problems
- problems in introducing new products
- poor delivery performance
- poor quality control
- high scrap levels
- poor production control

Externally, firms are being forced to respond much more to customer demand. This means shorter lead times on delivery, better delivery performance, high and consistent quality, and particularly, the ability to meet an increasingly customised product specification. With many markets becoming much more fashion-like with shortening product lifecycles, this puts severe pressure on firms to develop more flexibility and responsiveness characteristics which have been termed "manufacturing agility".

In response to these two pressures firms in the UK have tended to resort to technological answers as illustrated by Figure 1.

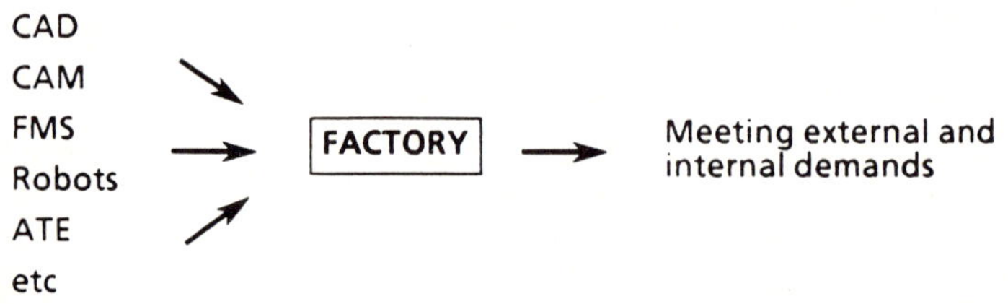

Figure 1

However, in common with other people (2) (3), we feel that this is an approach which fails to take note of both structural and methodological changes which can also effect an improvement, and which are occuring in other countries - notably Japan. These alternatives are illustrated in Figure 2.

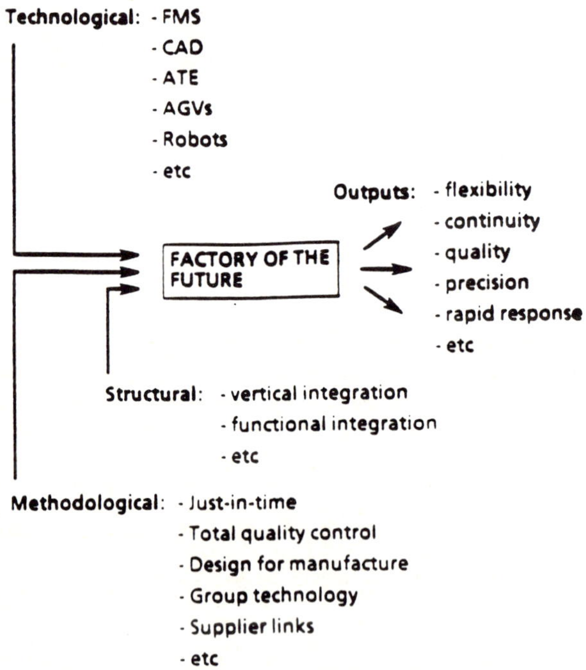

Figure 2

The Technology Option

The introduction of computers in the UK - as in all other countries - has tended to be rather fragmented in the past. There was little of it, it was expensive, it was physically energy and space consuming, there were few people who knew how to run it effectively, and it was devoted to specific functions such as salaries or tool stocks. All these factors are now rapidly changing and the integration of all aspects of manufacturing is growing from the receipt of orders through design, planning, production, assembly, despatch, etc. This can be visualised by reference to a model offered by Kaplinsky (4).

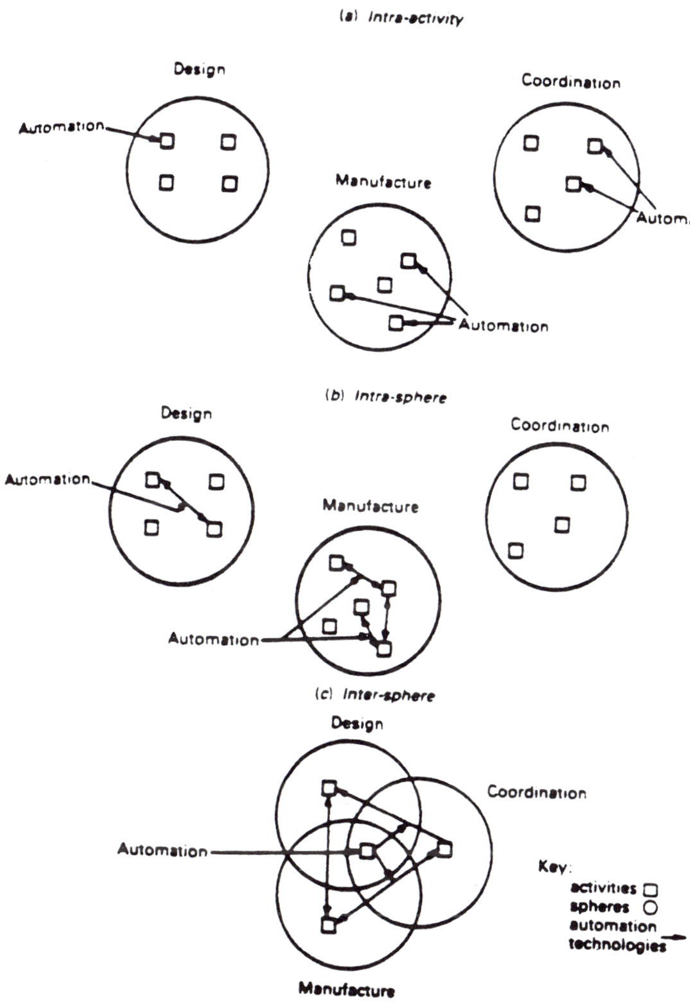

Figure 3

However, integration does not stop there, for increasingly companies in the UK are looking at links, both with their customers and their suppliers. In both cases they are examining their relations at the technical level where CAD-CAM can be so important, and the links that exist organisationally with other companies. In the latter case this is what Oliver Williamson (5) has called "obligational contracting". Richard Lamming (6) has observed that in the UK car industry, the traditional relationship between the buyer and the supplier has involved: closed competition; one-sided design effort; restricted information flow; and delivery patterns primarily geared to the buyer. He believes that the purchasing function will lose its overall control; or that a new "superbuyer" with all aspects of contact, and liaising between other specialists might emerge. He might perhaps be called a "procurement engineer". In these new terms therefore an expanded version of the Kaplinsky model is needed (7).

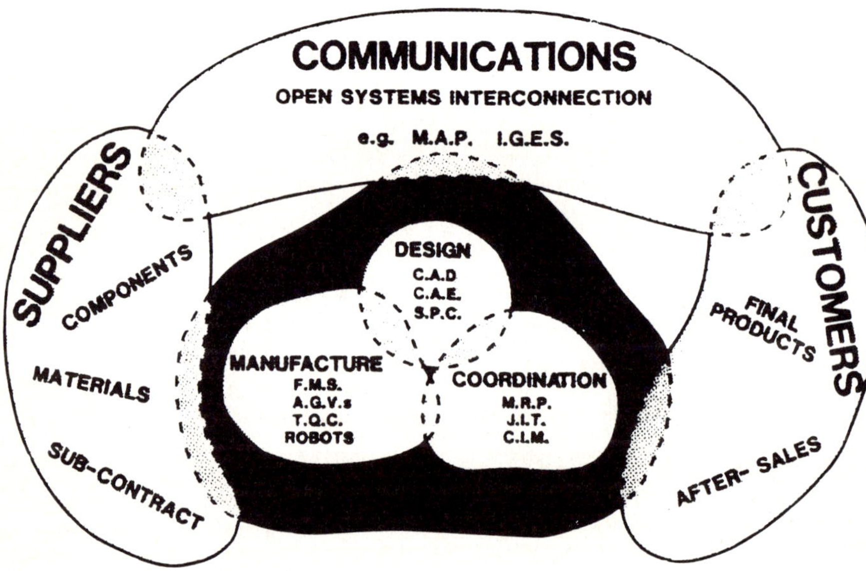

Figure 4

KEY	M.A.P.	Manufacturing Automation Protocol
	I.G.E.S.	Initial Graphics Exchange Specification
	C.A.D.	Computer Aided Design
	C.A.E.	Computer Aided Engineering
	S.P.C.	Statistical Process Control
	F.M.S.	Flexible Manufacturing Systems
	A.G.V.s	Automatic Guided Vehicles
	T.Q.C.	Total Quality Control
	M.R.P.	Material Requirements Planning
	J.I.T.	Just-in-Time
	C.I.M.	Computer Integrated Manufacturing

This offers one way in which these issues might be considered. For example, while Kaplinsky offers us a picture of integration within the firm, we anticipate growing links between firms as well. If this is the case then there will also be a growing need to establish an overall communications system which would allow common standards within Open Systems Interconnection networks; and examples of such developments would be the current General Motors initiative - Manufacturing Automated Protocols (M.A.P) - and the Initial Graphics Exchange Specification (I.G.E.S).

At a technological level the main questions concern getting the different hardware and software systems to work together. In our sample firms, most of the firms interviewed felt that software represented the major technical problem. Hardware problems mainly revolve around linking machine tools with handling systems and, since these technological developments have been evolving over a long period, they appear to pose few insurmountable difficulties. On the software side the picture is much more fluid with rapid proliferation of operating systems, communications options and control programs. The key issue is one of software integration - of providing structures and standards within which communication can take place on and between different levels in a control hierarchy. Problems are compounded with higher levels of integration - such as systems which involve some form of CAD-link.

The Need for Organisational Change

In a growing number of case studies it is becoming clear of the need for organisational adaptation. Simply deciding to invest in CIM is not in itself likely to bring the full range of potential benefits to the organisation. There has to be a degree of organisational change - development of skills, adoption of new working arrangements, changes in physical layout, changes in managerial structures, etc, - in order to obtain the full benefits. This adaptive process may take several years to complete. It can be seen as an extension or addition to the traditional learning curve which is normally defined only in terms of absorbing technical competence rather than including organisational development.

In our own research programme over the last three years - with some seventy companies in several countries and comprising in-depth interviews, literature analysis and questionnaires - a number of interesting aspects emerged in relation to the use of integrated technologies. In almost every interview we carried out, the importance of methodological routes to improved efficiency were stressed.

When accounting for overall improvements in efficiency, estimates of the importance of organisational change varied between 40% and 70% of the total improvement. Clearly around half of the improvement tended to come from these types of change. One company, which had not introduced FMS/FMC, had reduced the level of stocks held by 60% and increased output by 30% just by adopting "good management and methodological practices". Over the period of methodological change and subsequent installation of CNC standalone equipment the turnover of the company was trebled from £20m to £60m (1981-4). This was during a period when employment declined on average by over 30% in many other companies in the study yet employment grew in this non-FMS company by some 14%. In other words what was important here was not so much highly sophisticated integrated technology but the adoption of good management and production practices.

The rethink required encompasses both the hardware and the software elements of production, right through to the organisationally efficient management of production. In some instances simply conducting a feasability study into the implications of FMS had been enough to convince companies that substantial improvements could be made to efficiency by an overhaul of their production processes. Thus it is important to consider organisational and technological change options in approaching some concept of the "factory of the future".

It is becoming increasingly clear that such factors as: Just in Time Production; Total Quality Control; Supplier and Customer relationships; etc, are all extremely important in improving efficiency (8) (9), and that it is not merely the technology that we should consider. Significantly, it is precisely in these areas of organisational and methodological techniques that the Japanese are strongest; and which are somewhat belatedly attracting the attention of the West (10) (11).

While cultural reasons may be advanced why such "Japanese" style methodological developments are inappropriate for Western countries, it would be difficult to think of many "Japanese" production methods which do not have their roots in Western ideas, taken and developed by the Japanese in the 1950's and 60's. Theorists such as Sloan, and particularly Deming and Juran, have been extraordinarily important to the Japanese, and as Schonberger has pointed out:

> "The Japanese have had little trouble learning our techniques, and we will have little trouble learning theirs."

Manpower and Skills

In manpower and skill terms, at the level of work organisation, the two key trends which are apparent in the move towards flexible manufacturing are the overall reduction in the total number of workers required and the concurrent increase in both the breadth and depth of skill needed to support such systems (12). In particular, emphasis is shifting towards more direct activity, mainly concerned with maintenance and supervision of otherwise automated processes; in this batch manufacturing plants are beginning to resemble the typle of plant found in the process industries such as petroleum refining, where a small shift team is responsible for supervising and maintaining a large, highly automated plant. The range of skills required within, for example FMS teams, is wide and considerable emphasis is being placed on multi-skilling and the ability to switch rapidly between tasks; in addition, there is a growing need for new skills such as diagnosis, replacing much direct intervention.

Once again, the pattern is essentially one of integration between skills and between tasks associated with the manufacturing process. This implies a reversal of the traditional pattern in manufacturing which, following Taylor's Scientific Management approaches, was to break tasks down into small units; the trend now is away from such differentiation and towards a regeneration and broadening of skills. It is possible that this may set the pattern for the future of manufacturing employment, and perhaps we might suggest that Computer and Human Integrated Manufacturing (CHIM), is more appropriate than merely CIM.

It is also clear that one of the major constraints on investment in CIM has been the lack of in-house skills and knowledge. The recognition of severe shortages of multi-skilled personnel has been keenly felt by firms throughout the UK - and in other countries - with particular shortfalls of skilled engineers (both hardware and software), systems analysts, systems engineers, and good electrical/electronic engineers with "hands-on" experience.

Many of the skill requirements can be directly associated with the fact that the systems are becoming more complex and integrated. With the number of, previously discrete operations, now being completed within a single complex cell or system, this has meant a shift in the balance of direct to indirect, with those remaining direct workers bearing greater responsibility for a much more sophisticated and concentrated system.

The shift to integrated manufacture personified by FMS's also sharply brings into question the traditional pattern of functional specialisation. One of the most obvious changes necessitated (and indeed facilitated by technological change), is the merging of the design and manufacturing activities. The concept of "design for manufacture" is of particular significance for FMS where small modifications to the design of an item can eliminate the need for complex manipulation or operations within an automated system.

The essence of any such changes in functional specialisation is not to eliminate any of the specialist skills, but to assure that they are operating in a coordinated fashion. This requires the adoption of a single systems view of the process and the scrapping of the all too familiar parochial boundaries between departments.

The type of organisational changes required by integrating technologies do not stop at linkages between departments. Just as important as functional integration is vertical integration and shortened hierarchies. Most of the benefits associated with CIM and FMS come from the ability to respond in a rapid and flexible manner to new orders and changes in the operating parameters. Unless the technology is matched by a management decision-making structure which is closely involved with the shop floor and which has a high degree of shop-floor autonomy, the potential advantages of the technology will not be fulfilled. By shortening management hierarchies a firm can increase the effectiveness of their communications by reducing the time necessary to take and convey a decision.

SUMMARY

In many firms internationally, attempts are being made to mix the two strands of thought outlined above (the technological and the organisational), and to adopt, where appropriate, CIM while also attempting to improve other aspects of non-technical organisational change. It is possible to identify a simple matrix on which to locate the options for moving towards a fully computer-integrated factory (see Figure 5), from what is currently a dis-integrated form of production in both technological and organisational terms.

Increasing technological integration →

Increasing organisational integration ↓

Current state of factories - dis-integrated technology and organisation	High technological integration (eg CIM) but dis-integrated organisation
High organisational integration - the Japanese model	Integrated technology and organisation

Figure 5

Although the technology first route is more often applied, experience suggests that users of CIM technologies are becoming increasingly aware of the need for organisational change to support the technical. From this it is inferred that more viable alternative strategies would be to change the organisation first.

REFERENCES

1 J.R. Bessant, B.W. Haywood, Flexible Manufacturing Systems as an Example of Computer Integrated Manufacturing. Innovation Research Group, Occasional Paper No 1, Brighton, UK, 1985.

2 Ingersoll Engineers, Integrated Manufacture, I.F.S. Publications, Bedford, UK, 1985.

3 Professor J. Parnaby, Competitiveness Through Systems Engineering, Cambridge Manufacturing Forum, Cambridge, UK, 1986.

4 R. Kaplinksy, Automation : The Technology and Society, Longmans, London, UK, 1984.

5 O.E. Williamson, Transaction Time Economics, Journal of Law and Economics, Volume 22, No ii, 1979.

6 R. Lamming, Technical Change and Buyer - Supplier Relationships. Int' Jour' of Production and Operations Management, Volume 6, No 5, 1986.

7 R. Kaplinsky, op cit.

8 R.J. Schonberger, Japanese Manufacturing Techniques, The Free Press, New York, USA, 1982.

9 R. Dore, Goodwill and the Spirit of Market Capitalism, Jour' of Sociology, Volume 24, No 4, London, UK, 1983.

10 Just-in-Time : On the move and out of the Aisles, Production Engineer Magazine, London, UK, June 1984.

11 How Australian Industry Points the Way on Kanban, Production Engineer Magazine, London, UK, June 1984.

12 J.R. Bessant and P Senker, Societal Implications of New Manufacturing Technology, in: N Kemp, et al (Eds), The Human Side of New Manufacturing Technology, Wiley, Chichester, UK, 1986.

LATE SUBMISSION

Innovation: Adaptation and Growth, edited by R. Rothwell and J. Bessant
Elsevier Science Publishers B.V., Amsterdam, 1987 — Printed in The Netherlands

OAKLAND UNIVERSITY -- LINKS WITH THE OAKLAND TECHNOLOGY PARK: A PROACTIVE STRATEGY THAT IS WORKING

Frank P. Cardimen, Jr.
Director
Center for Economic Development
and Corporate Services
Oakland University

SUMMARY: While other universities and colleges in the United States are attempting to establish research parks, Oakland University is already well-positioned to take advantage of technological and economic growth in southeastern Michigan because of its unique relationship to the Oakland Technology Park. As a result, Oakland University is a leader in the changing attitudes among business and higher education in the United States.

Oakland University began its proactive strategy in 1981 by helping to establish the Oakland Technology Park on an 1,100 acre parcel of land adjacent to the university. This partnership is unique among universities affiliated with research parks because Oakland University has played a key role in the inception and economic development of the park although it does not own the land, nor is it charged with development or management. Oakland University has, instead, been the catalyst, bringing together the key partners of business, government and the community in what is destined to be the sixth or seventh largest research park in the United States.

INTRODUCTION -- UNIVERSITY-RELATED RESEARCH PARKS

In today's highly competitive world marketplace, aggressive organizations are constantly looking for new developments, improved products, better ways to serve customers--and employees who are well qualified and productive.

The functions of universities include education, research, and service--the keys to helping public and private enterprises succeed. Those have never been as valuable as they are today, in an information age in which technology expands our horizons every minute.

As a result of these phenomenal technological changes, businesses and universities in the United States are changing their attitudes toward each other. They are seeking ways to develop joint relationships that will speed the developmental transition from those products in the marketplace today to the new products of tomorrow. Reducing the time needed for technology transfer--the process of moving new developments from the research lab to industry and then back to higher education for refinement--is of vital concern to both competitive businesses and university researchers.

Technology transfer is just one reason why public and private universities are encouraging businesses to locate their research and development facilities in university-related research parks. A park site close to an academic institution breaks down spatial barriers and may thereby make it easier for academic and industrial researchers to interact more frequently and intensely, share each other's facilities and develop cooperative programs. Universities also are finding public and private funding increasingly inadequate for obtaining the technological resources needed to educate students for the future and to support graduate research. Equally important, businesses know the need for "brain gain" is essential if they are to keep their competitive edge in a world-wide marketplace.

U.S. Technology Parks

This phenomenon is certainly not new. The most successful parks in the United States to date--Stanford's Research Park (at Stanford University), North Carolina's Research Triangle Park (connecting the University of North Carolina, Duke University and North Carolina State University), and Boston's Route 128--were conceived in the mid 1950s. Each has spawned new companies and attracted governmental and corporate research and development facilities, and facilitated the transfer of new products to the marketplace.

Yet, the park site model was treated with caution--if not disapproval--as recently as a few years ago. Several studies in the 1960s and 1970s reported that more research parks were failing than succeeding.(ref. 1)

However, in the past few years there has been a resurgence of universities and colleges attempting to connect on the research park idea. One of the major reasons for the resurgence is the changing attitude between business and education--one of mutual dependence to survive the technological revolution currently taking place.

Appendix I lists the research parks in the United States today based on data from the Association on University Related Research Parks (AURRP). This international organization began two years ago because of the enormous activity by universities and colleges who were establishing or wanted to establish a research park on their vacant land.

More than 85 U.S. universities and colleges associated with the AURRP have now established research park affiliations. Of the 60 institutions responding to a recent questionnaire, 70% began after 1981 and 40% began after 1984. Of those launched after 1981, only 7.6% of the acreage allotted for their parks is occupied today. Of those parks that were begun before 1981 (30%), 44.2% of their land is occupied.

Sixty U.S. Research Parks
January 1, 1987

Year Initiated	No. Parks/ Universities	Total Acreage	Acreage Occupied	% Occupied
Prior to 1981	18	14,904	6,595	44.2%
1981 - 1987	42	24,830	1,890	7.6%

With only 44.2% of the land occupied in research parks that began prior to 1981, attracting "high tech" businesses is a long-term process.

Since 1981 at least 62 universities and colleges (42 reported plus 20 unreported) felt strong about investing land and dollars in the research park concept. To date less than 50% of these same parks have progressed beyond the conceptual or developmental stage. With only 7.6% of the land occupied in the parks, it appears that attracting high tech businesses to the research park will continue to be a long-term project.

Are there too many research parks and not enough "high tech" companies? Will businesses continue to be cautious before relocating? Will research parks prosper? It is still too early to determine success or failure, but the next few years will be critical for many of these university-related parks across the United States.

Whether or not this research park concept actually represents "the most dramatic contribution to innovation" of the various forms of technology transfer, it is the most visible. Precisely because of its concreteness, it is often the image that university presidents and trustees have in mind when thinking about academic/industrial cooperation and why many universities are converting their own land into research parks.(ref. 3)

What will it take for the new parks to succeed? What do businesses want when they choose to relocate to a university-related research park?

In a 1982 survey, nearly all responding tech park companies indicated access to college graduates (95.9%) among the most important factor in choosing to locate near a university of college. Two-thirds of the companies named recruitment of students as the primary method of tripping the mechanism of technology transfer. Students who are familiar with up-to-the-minute technology can help speed that knowledge from the lab to the workplace.

Technology transfer mechanisms from universities to high-technology companies	
Recruiting Students	66.9%
University publications (books, articles, etc.)	46.1%
Faculty consulting	41.8%
University services (seminars, conferences)	32.9%
Government dissemination of the results of basic research	30.5%
Corporate support for basic research at universities	25.1%

(Responses indicate attributes described as important or very important.) Source: Robert Premus, Location of High-Technology Firms and Regional Economic Development (Washington, D.C.: U.S. Congress, Joint Economic Committee, 1982).

Ninety-two percent (92%) of the companies look for increased opportunities for their employees to enroll in degree programs, access to libraries, information systems, and faculty research and consulting are other attributes sought by high-technology companies.

University attributes sought by high-technology companies	
College graduates	95.9%
Degree programs for employees	92.1%
Access to libraries and information systems	85.5%
Cultural activities	69.0%
Faculty consultants	66.7%
Faculty research activity	57.8%
Access to laboratories	52.3%
Part-time teaching opportunities for employees	22.3%

(These data include only companies that indicated physical proximity to a university system affected their choice of location. Responses indicate attributes described as most important or important.) Source: Robert Premus, Location of High-Technology Firms and Regional Economic Development (Washington, D.C.: U.S. Congress, Joint Economic Committee, 1982).

OAKLAND UNIVERSITY -- A PROACTIVE STRATEGY

Oakland University, a public university in southeastern Michigan, has 12,700 undergraduate and graduate students and a faculty that totals 600. It is among those universities that began exploring the concept of creating a technology park on vacant land surrounding its campus after 1981. The large amount of available land and such factors as convenient access to water, expressways and a nearby major metropolitan area--Detroit-- made the location promising for large scale development.

The university submitted its concept in 1982 to the Michigan High Technology Task Force (appointed by the governor), which was working to identify several Michigan sites for possible tech park development. The task force unanimously endorsed Oakland University's concept, making it the only one in the state to carry official endorsement. Associated with this endorsement, tenants receive a 12-year tax benefit that Oakland University successfully promoted at the state capitol in Lansing.

Thus, the missing ingredients in the development of this prime acreage adjacent to Oakland University were satisfied--initiative and tax incentives--all promoted actively by President Joseph E. Champagne.

Even while the park's concept was being created, Comerica Incorporated (second largest bank in Michigan) gave the project impetus by deciding to locate its new computer operations center there; Comerica broke ground in June 1983. When it opened two years later, Comerica's $42 million structure became the park's first occupied building.

A development consortium was formed among the park's landowners Comerica, Frankel Associates, Schostak Brothers & Co., Inc., Oakland University and Oakland Community College (a technical college specializing in high-technology education which also adjoined the park property). Preliminary plans for the park were developed under the direction of the consortium. Economic and market consultants also were retained. With marketing under the leadership of Schostak Brothers, the park began to take shape and the park's plans were officially announced in March 1984.

The resulting partnership is unique among universities affiliated with technology parks in the U.S. "We're educators, not land developers or marketing experts," said President Champagne. While Oakland University plays a key role in this development, the university does not own the land, nor is it charged with developing or marketing the park. Oakland University has been the <u>catalyst</u> in bringing together the partners in one of the few successful U.S. parks in the past few years. The Oakland Technology Park is destined to be the sixth or seventh largest technology park in the United States in size and investment dollars by 1992.(ref. 4)

To implement Oakland's new strategy, the university established the new Center for Economic Development and Corporate Services in 1985 to serve as the liaison between the park and the university community.

<u>The Oakland Technology Park</u>

The Oakland Technology Park is now an 1100-acre (445.3 hectares) research and development technology park located west of Rochester, Michigan (see Appendix II), in the cities of Rochester Hills and Auburn Hills. The park borders Oakland University's 1500-acre campus. It is situated 30 miles north of Detroit, Michigan. The park is being developed for use by organizations specializing in computers, robotics, applied engineering, automation and advanced manufacturing techniques, especially in relation to the automotive industry. There will be no manufacturing in this park.

Facilities will include an automotive technology center, computer operations centers, biotechnology and communications research centers, plastics R/D, regional and national headquarters of several firms, labor/management training facilities, a hotel complex and conference facilities, etc. (See Appendix III of companies currently located in the park.) Incubator space, which would provide entrepreneurs with inexpensive facilities to use while developing promising new products, also is planned.

More than 25,000 employees will work at the Oakland Technology Park upon its completion in the early 1990s with an investment of nearly $2.0 billion. As many as 30,000 additional jobs will be created within five miles of the park as a result of its magnetic force. This will bring an additional $1.0 billion investment to the area.(ref. 5)

Natural woodlands cover much of the 1,100-acre technology park, and the Clinton River meanders through the area.

Its rural character notwithstanding, the park is easily accessible to I-75, Michigan's main north-south highway, I-94, I-96, I-69 and M-59, a major east-west artery of the southeastern part of the state. It is within convenient driving distance of two airports, Pontiac-Oakland and Detroit Metropolitan, as well as dozens of cultural, recreation and entertainment enterprises.

Conclusion

Oakland University's involvement in the Oakland Technology Park is a unique model among U.S. universities today. The university has taken a proactive position in developing a technology transfer environment for its community, state, and country. The university, with its strong undergraduate and graduate programs, is stimulating growth for its students and faculty, is meeting business needs, and is forcefully creating opportunities to help shape technological and economic growth for southeastern Michigan.(ref. 6) The park also provides exciting opportunities to blend humanity with technology--and to restate Oakland University's fundamental role in preparing students for a complex and changing world.

Appendix I 1/3 UNITED STATES RESEARCH PARKS

Research Park	Date Initiated	Acreage Net	Acreage Occupied
**Arizona			
Arizona State University Research Park	1983	323	15
**California			
Stanford Research Park	1955	660	660
University Center for Science & Tech. Transfer	1984	400	0
**Colorado			
University of Colorado Research Park	1984	200	122
**Connecticut			
Connecticut Technology Park	1986	390	0
New Haven Science Park	1982	80	65
**Delaware			
University of Delaware Research Park at Lewes	1985(E)	107	0
**Florida			
Central Florida Research Park	1979	1094	197
Innovation Park - Fla. State & Fla. A&M	1979	208	8
Progress Center - University of Fla.	1984	1800	200
Tampa Technology Park	1985	1756	0
Univ. of Florida Research & Technology Park	1984	2200	260
**Georgia			
Advanced Technology Development Center	1980	3	3
University of Georgia Research Park	1980	400	140
**Hawaii			
Hawaii Ocean Science and Technology Park	1981	256	0
**Idaho			
Idaho State University Research & Business Park	1985	77	0
**Illinois			
Chicago Technology Park - Univ. of Illinois	1982	56	7
Evanston Univ. Res. Park - Northwestern Univ.	1984	26	5
Public Affairs Res. Ctr. - Sangamon State Univ.	1983	30	0
**Indiana			
Purdue Industrial Research Park	1980	177	140
**Kansas			
Univ. Corporate Research Park - Univ. of Kansas	1985	650	0

Appendix I 2/3 UNITED STATES RESEARCH PARKS

Research Park	Date Initiated	Acreage Net	Acreage Occupied
**Kentucky			
Northern Kentucky Research & Tech. Foundation	1980	75	9
Thomas More Center	1983	240	5
**Maryland			
Maryland Sci. & Tech. Ctr. - Univ. of Maryland	1984(E)	320	0
**Massachusetts			
Simplex Development Area	1985(E)	27	0
**Michigan			
Ann Arbor Tech. Park - University of Michigan	1982	820	96
Geddes (Huron) Center	1984(E)	508	0
Oakland Technology Park - Oakland University	**1983**	**1100**	**779**
**Missouri			
Missouri Research Park	1984	740	0
University of Missouri Columbia Research Park	1984	80	0
University of Missouri Kansas City Res. Park	1984(E)	65	0
**Nevada			
Dandini Research Park - University of Nevada	1986	470	0
**New Hampshire			
University of New Hampshire Technology Park	1986	835	0
**New Jersey			
Princeton Forrestal Center	1975	1750	1604
Princetonpark Corporate Center	1982	635	6
**New Mexico			
New Mexico State Univ. Research Park	1984	250	0
University of New Mexico Research Park	1963	104	15
**New York			
Cornell Industry Research Park	1965	200	180
Metrotech - Polytechnic Inst. of New York	1980	16	0
Rensselaer Technology Park - RPI	1979	1200	400
Rochester Institute of Tech. Res. & Dev. Park	1980	600	0
**North Carolina			
Research Triangle Park - U. of N.C., Duke U., and N.C. State U.	1956	6550	2704
University Res. Pk., Inc. - U. of N.C.-Charlotte	1980(E)	325	180

Appendix I 3/3 UNITED STATES RESEARCH PARKS

Research Park	Date Initiated	Acreage Net	Acreage Occupied
**Ohio			
Innovation Ctr. & Research Park - Ohio Univ.	1984	232	0
Miami Valley Res. Pk. - Dayton Central Wright St.	1980	1200	100
Ohio State University Research Complex	1983	700	1
**Oklahoma			
Swearingen Research Park - U. of Oklahoma	1982	600	150
**Oregon			
Riverfront Research Park - U. of Oregon	1984	71	0
**Pennsylvania			
Toftrees Corporate Park - Penn State	1984	300	0
University City Science Center (17 Institutions)	1964	17	5
**South Carolina			
Carolina Research Park - U. of South Carolina	1985	563	18
**Tennessee			
Tennessee Technology Cooridor - U. of Tennessee	1982	6100	117
**Texas			
Advanced Robotics Research Institute	1985(E)	18	0
Houston Area Research Center	1982	100	0
Texas A&M University Research Park	1982	434	5
**Utah			
University of Utah Research Park	1965	325	250
**Virginia			
Virginia Tech. Corporate Research Center - VPI	1985(E)	120	0
**Washington			
Washington State Univ. Research & Tech. Park	1982	156	6
**West Virginia			
Morgantown Industrial Research Park	1984(E)	670	20
**Wisconsin			
University of Wisconsin Research Park	1981	325	8

Appendix II

LOCATION OF OAKLAND TECHNOLOGY PARK
Michigan, U.S.A.

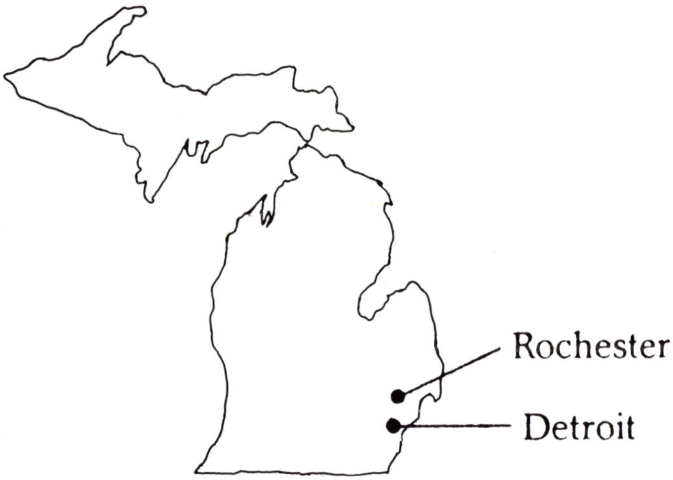

OAKLAND TECHNOLOGY PARK/OAKLAND UNIVERITY
Auburn Hills/Rochester, Michigan

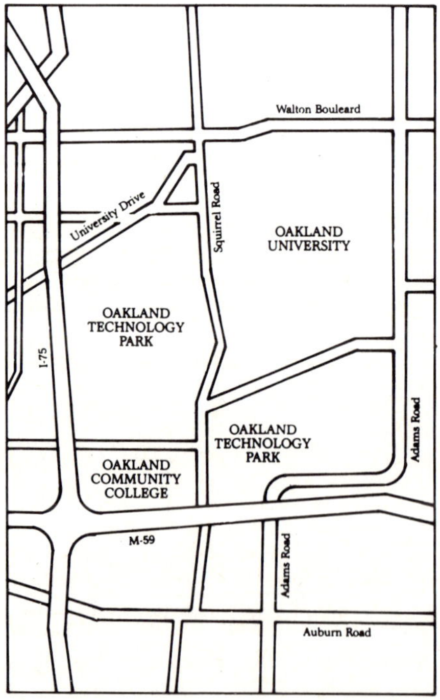

Appendix III

OAKLAND TECHNOLOGY PARK/OAKLAND UNIVERSITY
Auburn Hills, Michigan
Development Status March 1987

Company	Use	Square Feet	Employment
Cambridge Commons*	Hotel/Office Complex	32,000	150
Chrysler Technology** Center	Advanced Engineering/ Research Development	2,500,000	6-10,000
Comerica Incorporated	Bank Data Center	400,000	1,500
Electronic Data Systems (EDS) I and II	Data Processing Center and Offices	75,000 150,000	175 250
GKN, Inc.	U.S. Headquarters -- Auto Suspension	110,000	150
GMF Robotics*	World Headquarters Research/Development	200,000	450
Oak Tec Office Center	Offices	230,000	900
SecureData, Inc.*	Data Processing	50,000	170
World Computer	World Headquarters Research/Development	55,000	300

*Operational by June 1, 1987
**Ground broken October, 1986; first building by 1988; construction to 1991

REFERENCES

1 L.G. Johnson, The High-Technology Connection: Academic/Industrial Cooperation for Economic Growth, ASHE-ERIC Higher Education Research Reports, 1984, p. 63.

2 Association of University Related Research Parks, Data Compiled at January 1, 1987, Arizona State University, Tempe, AZ.

3. General Accounting Office, Joint Economic Committee, 1983.

4. Association of University Related Research Parks, Data Compiled at January 1, 1987, Arizona State University, Tempe, AZ.

5. Oakland County Economic Development Group's Forecast.

6. J. Main, Fortune, "...Each professional researcher adds about $125,000 a year to a local economy.", "Business Goes to College for Brain Gain", March 16, 1986, p. 84.